access to geography

WEATHERING, SLOPES *and* LANDFORMS

David Atkinson

Hodder & Stoughton

A MEMBER OF THE HODDER HEADLINE GROUP

For Clare

The author would like to thank his colleagues at Dr Challoner's Grammar School, Amersham for their support and friendship.

Acknowledgements
The publishers would like to thank the following individuals, institutions and companies for permission to reproduce copyright illustrations in this book:

M.J. Clark and R.J. Small, *Slopes and Weathering*, 1982 © Cambridge University Press, reprinted with permission of Cambridge University Press, pages 100, 104, 105, 107; A. Dolgoff, *Physical Geology*, 1996 © D C Heath, reproduced by permission of Houghton Mifflin Company, pages 95, 118, 119; Michael Hill, pages 106, 124; D. Edwards and C. King, *Geoscience*, Hodder and Stoughton, 1999, reproduced by permission of Hodder Arnold, pages 3, 21; Jacques Langevin/CORBIS SYGMA, page 92; Danny Lehman/CORBIS, page 23; Stan Pritchard/Photographers Direct, page 6; Joseph Sohm, ChromoSohm Inc/CORBIS, page 74; M.A. Summerfield, *Global Geomorphology: An Introduction to the Study of Landforms*, 1991, Pearson Education Limited, page 42. All other photos by the author.

Every effort has been made to trace and acknowledge ownership of copyright. The publishers will be glad to make suitable arrangements with any copyright holders whom it has not been possible to contact.

Note about the Internet links in the book. The user should be aware that URLs or web addresses change regularly. Every effort has been made to ensure the accuracy of the URLs provided in this book on going to press. It is inevitable, however, that some will change. It is sometimes possible to find a relocated web page, by just typing in the address of the home page for a website in the URL window of your browser.

Orders: please contact Bookpoint Ltd, 130 Milton Park, Abingdon, Oxon OX14 4SB. Telephone: (44) 01235 827720. Fax: (44) 01235 400454. Lines are open from 9.00 to 6.00, Monday to Saturday, with a 24 hour message answering service. You can also order through our website www.hodderheadline.co.uk.

British Library Cataloguing in Publication Data
A catalogue record for this title is available from the British Library

ISBN 0 340 81690 2

First Published 2004
Impression number 10 9 8 7 6 5 4 3 2 1
Year 2010 2009 2008 2007 2006 2005 2004

Cover photo: Aguelzim, in the Moroccan High Atlas (by David Atkinson).

Produced by Gray Publishing, Tunbridge Wells, Kent
Printed in Malta for Hodder & Stoughton Educational, a division of Hodder Headline, 338 Euston Road, London NW1 3BH.

Contents

Introduction

Geomorphology is the study of the shape of the surface of the Earth. It involves examining the influence on the landscape of a variety of forces, such as running water, ice, marine action and wind. However, these forces have to act on the substance and structure of the surface of the Earth. Consequently, the features found on the Earth's surface owe as much to the rocks that make up that surface as to the processes acting on them. Often this element of the study of the landscape is overlooked.

Different rock types have very different origins. Igneous rocks, made from solidified magma or lava from the mantle, are the primary building blocks of the Earth's crust. As all rocks are attacked by the elements they break down into particles that may subsequently be compressed into sedimentary rocks. If any rock is subjected to the intense heat or pressure that can be experienced through tectonic activity or close exposure to mantle rocks, it may be altered chemically and physically to form a metamorphic rock.

Each of these groups of rock gives rise to distinctive landscapes as a result of their structure or composition. Igneous rocks can form huge upland areas if large quantities of magma have cooled, or intricate patterns of ridges if material has been forced between other rock layers. Sedimentary rocks can yield remarkable landscapes governed by the chemistry of the sediment. Metamorphic rocks often have little pattern, in keeping with the chaotic nature of their formation.

Geological processes of tension and compression also have an impact on the shape of the surface of the Earth, and at a range of scales. Tension will create valleys thousands of kilometres long, while compression will build mountain ranges of similar magnitude. Activity of this nature has impacts on humans. Fossil fuel deposits will accumulate in folded rocks, while the hazards associated with life in mountainous areas are significant.

The elements attack any rock that is exposed at the surface by a number of processes. Mechanical actions can break rocks apart through force, while chemical reactions can disintegrate the rock from within by removing part of the mineral structure of the rock. The effectiveness of these processes vary across the world, and produce very different results in different areas. Many of the products of weathering are as much a result of the nature of the rock as the weathering processes, emphasising the fact that relationships between different areas of study are strong.

Once material has been weathered it is likely to move, and mass movement processes are key to developing an understanding of the workings of the surface of the Earth. Material can move downhill by a number of methods, ranging from gentle creeps to catastrophic slides and falls. A major landslide can devastate an area, causing significant loss of life.

All mass movement processes take place on slopes that develop over a period of time. By combining an appreciation of how slopes change with an understanding of the processes that take place on them, it is possible to interpret the landscape and estimate where hazards are greatest.

All environments pose challenges and opportunities for human communities. Those environments that are unique to particular rock types are no exception, and settlement, agriculture, mineral extraction and tourism are all human processes that can rely on the natural environment as dictated by the rock type.

1 Classification of Rock Types

KEY WORDS

Rock cycle: the movement of material through space and time as it transfers and transforms from one type of rock in one location to other rock types and places.
Igneous rock: the primary rocks of the Earth, formed as molten material from the mantle cools down, allowing the minerals to crystallise.
Sedimentary rock: a rock created from debris from another source, either rock or organic. Rock is formed when the debris is subject to great pressure.
Metamorphic rock: a rock that has been exposed to physical conditions that have resulted in the breakdown of previously stable minerals and the growth of new ones.
Mineral: a chemical substance with a definite composition and internal structure. Igneous rocks are made from minerals; the nature of the rock is defined by the variety and quantity of different minerals.
Crystallisation: the formation of a solid crystal of a mineral from the molten state as a magma or lava cools. Different minerals crystallise at different temperatures.
Clasts: fragments of material produced by the breakdown of existing rocks. Clasts form 85% of sedimentary rocks.
Diagenesis: the group of processes that affect all sediments as they are buried underground and are exposed to higher temperatures and greater pressures.

1 The Rock Cycle

The **rock cycle** is a representation of the fundamental relationships between all the rocks that make up the Earth. It demonstrates the way in which they are linked together, and shows the importance of plate tectonics in driving the cycle. It was first expressed by James Hutton (1726–1797), a Scottish scientist widely regarded as the founder of modern geology.

To understand the rock cycle it is simplest to begin with the basic building blocks of the Earth, the **igneous rocks**, since all other rock types are derived from them in some way. Igneous rocks are crystallised from molten material as it cools. If the molten material is on the surface when it cools it is known as lava, and forms **extrusive**

rocks. If the molten material is underground it is known as magma, and forms **intrusive rocks**. As soon as rocks are exposed at the surface of the Earth they begin to be broken down by processes of **weathering** and **erosion**. Different rocks produce different end products when they are broken down. Granite, for example, will be broken down to clays and sands. These particles may then be transported by air, water or ice and deposited in an environment that encourages the formation of **sedimentary rock**, such as sandstone or mudstone.

Over a period of many millions of years the sedimentary rock may then be buried as more sediment is added on top. If any rock is buried or located near to a source of great heat or pressure, it may be changed, or metamorphosed, to create a **metamorphic rock**. A limestone, for example, may be metamorphosed into marble. If temperatures exceed 800°C some rocks will melt and then reform as a different igneous rock. Many metamorphic rocks form when they are subjected to plate movements that move the rocks to a location where these temperatures are to be found. The oceanic trench of a subduction zone or the roots of a mountain belt would be examples of such locations.

Figure 1.1 shows how these processes are intertwined, and how they begin with the Earth's mantle. The energy to drive the rock cycle is derived from the heat continuously produced by radioactive decay within the Earth. Without that energy, the movements of plate tectonics would not occur, and the changes of the rock cycle would be much reduced.

The rock cycle is inextricably joined to the other great cycles of natural science. The carbon cycle has significant links in that the greatest single store of carbon is in carbonate rocks, such as limestone. The processes of weathering that break down rocks on the surface thus have an important role to play in the carbon cycle, since carbon is released from rock by such processes. The hydrological cycle, meanwhile, provides many of the processes that are responsible for weathering and erosion through moving water, freeze–thaw action and chemical processes.

The rock cycle began in the **Archaean period** of geological time, in excess of 4 billion years ago, when the first solid rocks and oceans formed. The oldest rocks are thought to be in north-west Australia, dating from approximately 4600 million years ago, while the first oceans were probably formed some 3800 million years ago. Since then there has been a constant recycling of rock material, as material has been transferred from an igneous rock to a sedimentary one and then, perhaps, to a metamorphic rock.

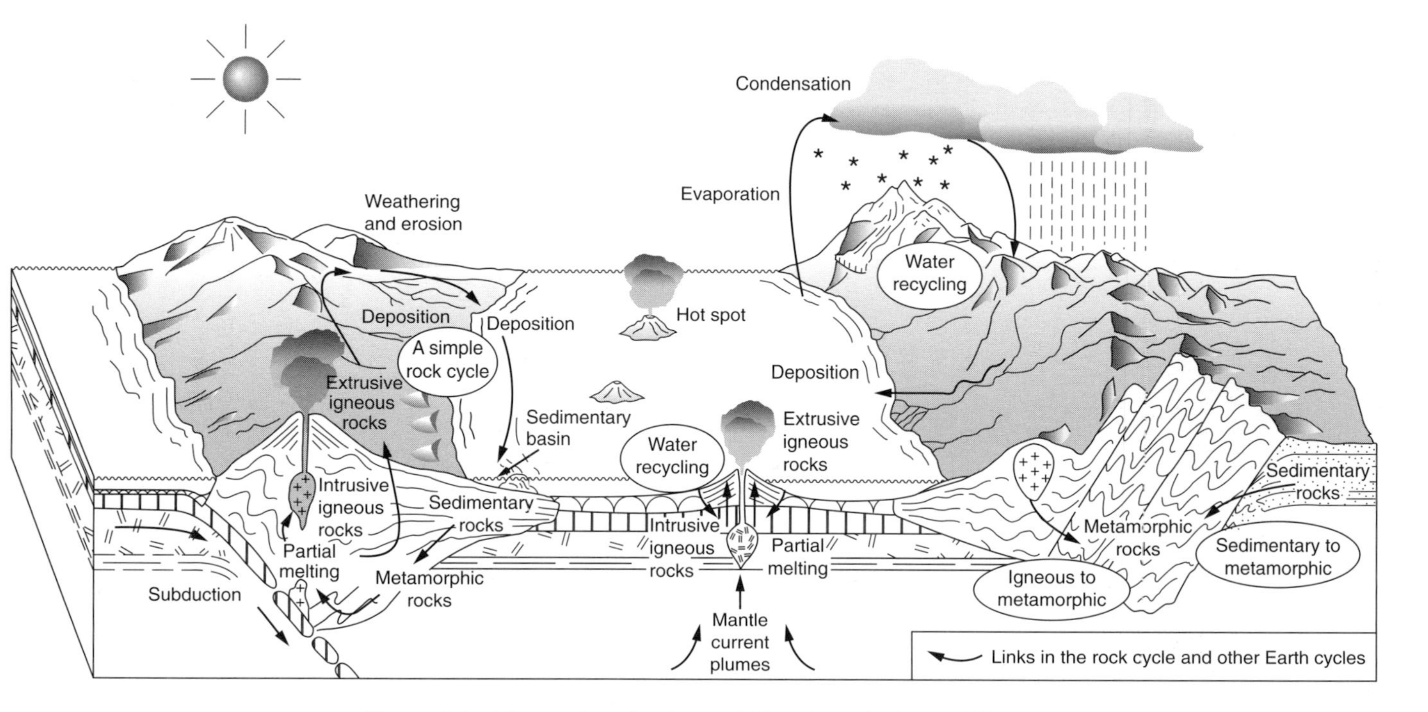

Figure 1.1 The rock cycle. *Source*: Edwards and King (1999).

2 Igneous Rocks

Igneous rocks are formed by the solidification of molten magma or lava from the mantle: they are the commonest rock type found on Earth. The oldest rocks yet dated are igneous and are to be found in Greenland, with an estimated age of 3900 million years old. However, there is a great range of igneous rock to be found, and a significant diversity in the processes of their creation. Igneous rocks can broadly be divided into two categories: **intrusive** and **extrusive**. Intrusive igneous rocks are those that formed under the surface, usually as material from the mantle forced its way through the crust, cooling and solidifying as it moved. Extrusive igneous rocks are those that formed above the surface, as lava erupted from volcanoes and cooled on contact with the atmosphere or the sea.

All igneous rocks have one thing in common in that they are made up of crystals of minerals, most commonly silicates. As molten magma or lava cools it is the **crystallisation** of these minerals that causes the rock to solidify. The nature of the rock is determined by the mineral content and the rate at which crystallisation occurred.

a) Formation and characteristics of intrusive igneous rocks

Intrusive igneous rocks are those that form as crystallisation occurs underground. They are sometimes referred to as **plutonic rocks**. Magma can be thought of as a group of minerals melted and mixed up together. As this group of minerals cools, the individual minerals will not solidify until their particular freezing point is reached. The molten masses that form granites, one of the most important igneous rocks, are usually created by the heat generated at destructive plate margins. This heat melts the lower layers of the crust, forming a magma that is thought to be rather viscous in nature. As a consequence of this viscosity the crystals of different minerals do not settle out as they form, but remain distinctly separate and interlock with crystals of other minerals. The interlocking crystal structure of **granite** can be seen in Figure 1.2.

There is a vast range of intrusive igneous rocks determined by the chemistry of the magma from which they were formed. **Acidic** rocks contain more than 66% **silica** and more than 10% **quartz** and are among the most common intrusive rocks, as are gabbros that are chemically basic with higher quantities of ferromagnesian minerals such as olivine and pyroxene. **Basic** rocks are quartz free, and contain 45–55% silica. The common ground between all intrusive igneous rocks is that they have large crystals that are clearly visible to the naked eye, as a consequence of their slow rate of cooling, crystallisation and formation underground.

South-west England provides a good example of intrusive igneous activity. There are many outcrops of granite throughout the penin-

Figure 1.2 Granite.

sula, such as Dartmoor, Bodmin Moor, Land's End and the Scilly Isles, and they are all offshoots of the same granite mass. Studies of the Dartmoor granite have shown that it cooled about 280 million years ago, some 6 km below the surface at the time. Subsequent erosion has brought it to the surface where it is now undergoing weathering and erosion as part of the rock cycle. **Gabbros** are found in highland areas of Scotland, perhaps most dramatically in the Black Cuillins of Skye (Figure 1.3) where the resistance of intrusive igneous rocks to weathering and erosion is clearly demonstrated.

b) Formation and characteristics of extrusive igneous rocks

Extrusive igneous rocks are those that form as crystallisation occurs above the surface of the Earth. They are sometimes referred to as volcanic rocks, a reference to the fact that they are often formed from lava that has erupted from a volcano, although they can also be produced along **fissures**, such as those found in Iceland. The nature of volcanoes is not part of this book, nor is the range of rock particles produced by volcanoes, but the nature of the rocks that form from lava is.

When lava is erupted from a volcano it immediately begins to cool as it comes into contact with air or, in the case of a submarine volcanic eruption, the sea. The chemical characteristics of the lava have an

Figure 1.3 Corrie Lagan, Black Cuillins, Isle of Skye.

impact on the rock produced, as does the rate of cooling of the lava. An acidic lava, similar to a magma that would generate a granite if cooled underground, will produce a **rhyolite** if it is exposed to the faster cooling processes that occur above ground. Equally, a chemically basic lava that would lead to the formation of a **gabbro** if formed intrusively will generate a **basalt** if it forms above ground. The silica content, as mentioned on page 4, is a significant factor in determining the type of extrusive rock formed. If the lava erupted is high in silica (an acidic lava) it will be a viscous and sticky lava because silica has a melting point in excess of 850°C. Consequently, it begins to solidify as soon as it comes into contact with air, causing the lava to flow only a short distance from the vent. This has an impact on the shape of the volcanic cone, creating a steep-sided feature, and the rocks in such cones tend to be rhyolites. If the lava has low levels of silica (a basic lava) it will have a lower melting point and consequently will flow further, creating plateaux or low-angle volcanic structures, such as the **shield volcanoes** found in Iceland or Hawaii. The rocks found in these features are usually basalts.

Although the chemical characteristics of these pairs of rocks (granite and rhyolite, gabbro and basalt) are the same, their appearances

Figure 1.4 Basalt.

are different. Their different rates of cooling mean that there are variations in crystal size between them. As the extrusive lava cools more quickly it forms smaller crystals than the slower cooling intrusive magma. Consequently, the crystals in extrusive rocks vary from the truly microscopic to basalts that can have crystals just about visible to the naked eye. Compare Figure 1.2 with Figure 1.4 to see the range of appearances that igneous rocks can adopt. Figure 1.2 is a granite showing clear individual crystals, while Figure 1.4 is a basalt in which the single crystals cannot be clearly seen, and the rock has the appearance of a solid structure.

3 The Formation and Characteristics of Sedimentary Rocks

As igneous and metamorphic rocks are exposed to the forces of weathering and erosion (which will be discussed in Chapter 4) they break down and disintegrate. These processes create innumerable particles of rocks and minerals, known as **clasts**, which can then be reworked, eventually, to form sedimentary rocks. This process accounts for 85% of sedimentary rocks and these rocks will take on many of the characteristics of the particles that make them up. Sedimentary rocks can also be created from organic material. The application of a substantial amount of **pressure** is needed in order to turn an accumulation of particles or organic material into a solid rock; although a range of chemical processes also contributes by holding particles together or changing the nature of the minerals in the clasts.

a) The formation of sedimentary rocks

In order for loose clasts or organic material to be turned into rock it is necessary that they be buried. Material moves deeper and deeper as more sediment is deposited above it. As this happens the material enters environments that are progressively hotter and subject to greater pressure. In turn, the sediment is subject to **diagenesis**, a set of processes that occur because the sediment is in an environment where it becomes **unstable**. Since all the elements of sedimentary rocks were stable at the surface where they were either part of another rock, or some form of organic structure, it is understandable that they will become unstable at depths in excess of 10 km and temperatures over 200°C. There is no clearly defined break between diagenesis and metamorphism of rock. As depth and heat increase, so the changes are more metamorphic in character. The best way to distinguish between them is to consider the nature of the minerals in the clasts. If the composition of those minerals changes then the process taking place is metamorphism. If the composition does not alter then the process is diagenesis. As the diagenetic processes take place they cause the clasts or organic material to be bound together and, eventually, turned into rock. Sedimentary rocks display different weaknesses, and therefore different diagenetic processes are significant in the formation of different rocks. Figure 1.5 shows some of the key processes involved in the formation of sedimentary rocks.

The type of rock produced is a result of the sediment that makes it up. Some rocks are constructed by a purely mechanical set of processes, while others have some chemical activity as part of their

Diagenetic process	Description
Compaction	Sediment is squeezed by the pressure of the overlying sediment; water is lost as porosity and permeability are reduced
Solution	Flowing pore waters have increased power to dissolve and transport material because of the high water pressures
Pressure solution	Minerals dissolve more readily at points where they are under pressure, so solution occur s where grains press into one another
Recrystallisation	Some minerals, particularly calcium carbonates, change in shape and size
Cementation	Minerals crystallise in pore spaces from circulating fluids, cementing grains together

Figure 1.5 Key processes involved in forming sedimentary rocks.

formation. The common mechanically formed rocks are **clays, marls** and **shales**, formed from clasts less than 0.062 mm in diameter; **sandstones** and **gritstones** (between 0.062 and 2 mm in diameter); and **breccias** and **conglomerates** (greater than 2 mm diameter). It is not uncommon to find some form of chemical cement holding the clasts together, especially in breccias or conglomerates. This chemical cement is precipitated out of the water into the gaps between the clasts.

The other significant group of sedimentary rock is those that arc formed from organic matter. As wide a range of rocks falls into this group as into the mechanical group. **Limestones** are produced from the remains of sea creatures that were subject to diagenetic processes on the bed of warm, calm seas. Most limestones are made mainly of calcium carbonate, and are known together as **calcareous** rocks. Peat and coal are produced as ancient woodlands decay. Peat forms as decomposition progresses and coal is the product of burial and associated increases in temperature and pressure.

Some sedimentary rocks are formed predominantly as a result of chemical processes. Evaporation of water leaving salts behind, or precipitation of salts around a central point can create rock. Flint, for example, is a consequence of the amalgamation of silica around a microscopic organism.

b) Environments favouring development of sedimentary rocks

Sedimentary rocks can form in a number of different environments. Many form beneath the sea, where the necessary mass of sediment can accumulate as rivers deposit huge quantities of material. In order to create the large areas of sedimentary rocks that can now be found around the world, very large areas of ocean floor must have been filled with sediment. These vast troughs are known as **sedimentary basins**, and our understanding of their existence owes a great deal to the theory of plate tectonics. Studies suggest that much sediment accumulates in the trenches, associated with tectonic subduction zones. This material is turned to rock, and is often reworked by subsequent tectonic activity. Equally, the basins created between island arcs and the associated mainland can create the conditions required for some sorts of diagenesis. On the modern map of the Earth's plates, the Sea of Japan, between China and Japan, is a good example of this, as particles are washed into the sheltered seawaters from the adjoining land. More simply, the shallow water margins of the modern Atlantic Ocean are being filled with sediment derived from the neighbouring land masses. This idea is shown on Figure 1.1. The opening of the Atlantic Ocean during the past 200 million years has created this space, illustrating the significance of plate tectonic activity in yet another aspect of geology. Also, in some places the crust is being stretched. As it stretches it becomes thinner and can sink, creating a depression. The North Sea has undergone this process and is consequently being filled with sediment from eastern Britain from one side and mainland Europe from the other.

These processes that are going on today also occurred in the past. Areas of warm, shallow tropical sea, such as the modern Caribbean, were filled with the remains of the abundant life forms of such a location to create thick layers of limestone.

Not all sedimentary rock is formed underwater. **Loess** is a rock-like material formed by the action of the wind. In arid zones wind carries particles more easily than if there is plentiful water, which encourages vegetation growth. These particles are subsequently deposited and can form layers of up to 250 m thick. Often the loess is trapped by vegetation, stabilised and compacted, going on to become the parent material for soil. Although the diagenetic processes discussed earlier will not occur because the thicknesses of sediment are insufficient, loess has sufficient cohesion, at least when dry, to be worthy of consideration alongside sedimentary rocks. However, when it is wet it loses much of that cohesion, and can be very vulnerable to landslides and slope failures.

c) The characteristics of sedimentary rocks

The key characteristics of sedimentary rocks can be grouped by three categories. The **chemistry**, **grain size** and **porosity** of the rock all contribute to the rock's resistance to weathering and erosion, and thus determine the landscape features produced.

The rocks that are made mainly of calcium carbonate have specific chemical properties that influence the landscape. Rainwater dissolves the calcium carbonate, meaning that the rock surfaces that are exposed to the rainwater are much more vulnerable to weathering and erosion. The mechanically formed sedimentary rocks tend to have chemical characteristics that are derived from the source material, rather than a set unique to the sedimentary rock. Sandstones, for example, are not vulnerable to chemical weathering because they are made predominantly of quartz crystals that are very resistant to such an attack.

In many mechanically formed sedimentary rocks the weakest element is the precipitated sediment that glues the clasts together. Thus, the larger the clast size, the more resistant the rock is likely to be. The size of clast also contributes to the porosity and permeability of the rock.

Porosity and **permeability** are critical elements in the weathering of sedimentary rocks because they control the ease with which water can gain access to and flow through the rock. **Primary porosity** is the amount of space within the rock structure: the ratio between solid material and pore space. Chalk has a very high primary porosity because there are many gaps between the grains in the rock. **Secondary porosity** is the extent to which water can pass through joints or fractures. In many limestones, secondary porosity is very high because of the bedding planes and joints in the rock, even though the primary porosity is very low. This has enormous significance for the landscapes that develop on limestone, as will be seen later.

Permeability is a measure of the rate at which fluids can pass through a rock. This is important to the study of landscape formation because the rate at which water can pass through a rock contributes to the rate at which processes of erosion and weathering can occur. If a rock has a fine-grained structure it could have a high primary porosity, but a low permeability if the pore spaces are so small as to prevent the rapid transfer of water through the rock. Clays are a good example of this. The high porosity but low permeability of clay increases the flood risk in such environments because the water cannot escape downwards at any speed.

4 The Formation and Characteristics of Metamorphic Rocks

Metamorphic rocks are, perhaps, the most diverse of the three rock families. Derived from a Greek word, metamorphic means changed in shape or form, and any rock that has its mineralogical composition or physical structure altered in some way becomes a metamorphic rock. The **forces** required to carry out the alteration are considerable, but the energy involved in the movement of plates is more than sufficient to metamorphose rocks on and under the surface. It is harder to classify metamorphic rocks than igneous or sedimentary because there are more variables involved. The original rock type will have a bearing on the product, as will the nature and amount of metamorphosis that occurs. Rocks that were subject to change very close to a plate boundary are likely to display more significant alterations than those that were further away from the source of the changes.

a) The formation of metamorphic rocks

Rocks metamorphose because they are made up of minerals. These minerals are vulnerable to changes in the same way that any household product is. If a sheet of plastic is placed in an oven it will melt. If a mineral is buried at great depth within the Earth it too will melt. The plastic is **stable** at room temperature, but becomes **unstable** as the heat increases in the oven. The mineral is stable at atmospheric temperature and pressure, but becomes unstable as the pressure and temperature increase as it is buried. The plastic changes its characteristics as it melts by changing its shape and giving off fumes, showing that a chemical reaction is occurring that alters its properties. As the mineral is buried it too will alter, although the alterations are to the mineralogical composition. Minerals do not melt like plastic, but their internal arrangement changes. The precise nature of these changes will depend on the mineral concerned and the forces applied to it, but the key point is that any mineral will change once it is subjected to temperatures and pressures outside its zone of stability.

Metamorphic activity can be divided into **contact metamorphism** and **regional metamorphism**. Contact metamorphism occurs when rock comes into contact with a source of extreme heat, such as an igneous intrusion. Regional metamorphism occurs when large areas are subject to the forces associated with a plate boundary, such as the world's great mountain ranges. These forces increase both temperature and pressure, while contact metamorphism tends to be caused by temperature increases only.

Contact metamorphism

Contact metamorphism, as with most metamorphic activity, occurs when the chemical content of the rock does not change, although the balance of minerals within the rock could be altered. This is known as an **isochemical** process. Water can be driven off by the increased temperature, and minerals can be recrystallised or broken down into their constituent parts. One of the better known metamorphic rocks is **marble**, widely used as building stone and for sculpture, such as Michelangelo's *David*. Marble is metamorphosed limestone and is, as such, a carbonate rock. When limestone is subject to the great temperatures (anywhere between 200 and 800°C) required for metamorphism, the calcium carbonate components decrystallise and then re-crystallise to form a different structure and appearance. Any fossils present in the limestone would have been destroyed, and the new rock would look utterly unlike limestone. This process can happen to any rock if it comes into contact with a source of great heat. It is common to find a **metamorphic aureole** surrounding an area that was such a source. If a major igneous intrusion occurs, such as the **batholiths** of south-west England, a band of rock surrounding it will be subject to contact metamorphism. As distance from the intrusion increases so the degree of alteration will decrease because the temperature and duration of the metamorphic processes will decrease. This helps explain the complexity of metamorphic rock classification. If the area surrounding an intrusion had originally been one rock type it would be possible to find a range of metamorphic rocks within that area according to the severity of the metamorphism experienced. That would, in turn, be related to the distance from the heat source. Thus a metamorphosed mudstone could be a **slate**, a **schist** or a **gneiss**, depending on the temperatures and pressures experienced by the source rock.

Regional metamorphism

Regional metamorphism occurs when plate boundary movements create conditions of high temperature and high pressure. The rocks present are thus subject to both metamorphic processes. Constructive plate margins produce the least dramatic metamorphoses, although it is thought that the crustal rocks close to such margins experience some low-grade changes.

However, it is at convergent plate margins that the most dramatic changes occur. A cursory glance at a geological map of the Himalaya or any other range of mountains will soon illustrate this. The dramatic colouring of such a map indicates a wide range of rock types, and the sharply defined edges to the blocks of rock show the great forces that have been applied. As the plates have moved together, the rocks have been distorted physically and altered chemically over a long period of time. There are rocks in the Alps, for instance, that show signs of metamorphism as long as 90 million years ago, and others that were altered as recently as 5 million years ago. While the results of metamorphism in a continental collision zone are clearly visible in mountain ranges, subduction zones also produce impressive features. Sedimentary rocks that form in the ocean trenches neighbouring a subduction zone can be subjected to great pressures as the plates converge, metamorphosing them into other types of rock. Equally, as the melted crustal material forces its way upward through the continental crust the heat and pressure can change the composition of the rocks it passes through.

Summary Diagram

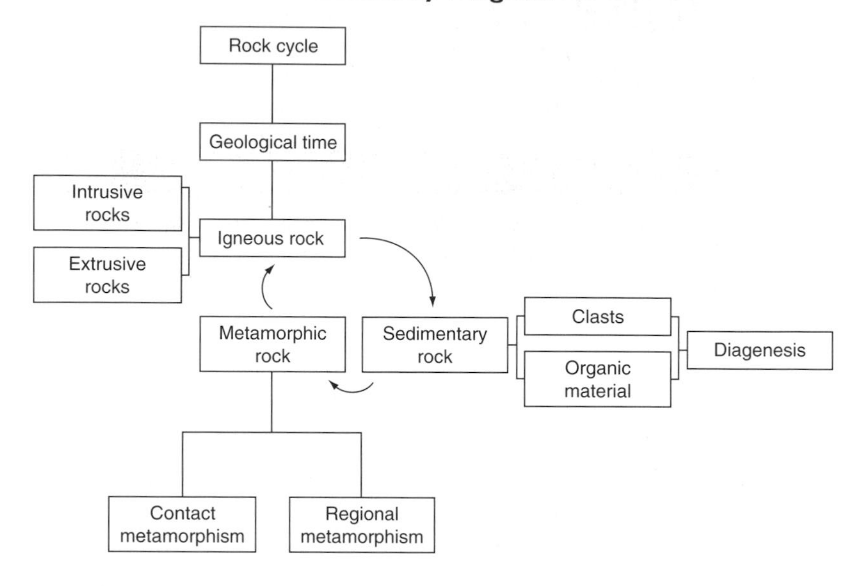

Questions

1. What conditions must exist for a planet to exhibit a rock cycle?
2. What are the main characteristics of:
   a) igneous rocks
   b) sedimentary rocks
   c) metamorphic rocks?
3. **a)** What is the relationship between cooling and crystal size in igneous rocks?
   b) Describe the differences that occur between intrusive and extrusive igneous rocks.
4. What sort of sedimentary rocks might you expect to develop in the following situations:
   a) on land, close to a mountain range
   b) on a continental shelf
   c) on the ocean floor, out to sea?
5. Describe the conditions likely to cause regional and contact metamorphism.

2 Landforms of Major Rock Types

KEY WORDS

Intrusive rocks: rocks that form when a body of magma is forced into existing crustal rocks and then cools and solidifies.
Extrusive rocks: rocks that form when lava has been ejected from a volcano and then cool and solidify.
Batholith: a large body of intrusive igneous rock that has no discernible base.
Karst: terrain created on limestone, with almost no surface drainage and a series of surface depressions and hollows.
Tor: a small exposure of well-jointed blocks of rock, usually on the top of a hill and often surrounded by a mass of collapsed blocks.
Cuesta: an asymmetrical ridge produced by differential erosion, with a gentle dip slope and a steep scarp slope. Cuestas are also known as **escarpments**.

1 Influences on Landscape

The rock types discussed in Chapter 1 have characteristics that can lead to a range of different results. Characteristics of **chemistry**, **resistance** to weathering and erosion, **permeability** and **structure** combine to create landscapes that could be said to be confined to that particular rock. However, some features appear on a variety of different rock types, although the method by which they are created can vary.

Landscapes are a product of their underlying geology, and as such are influenced by all the characteristics of that geology, from the purely physical such as the joints and bedding plane structures common in limestones, to the presence of particular minerals, such as feldspar in some granites, that contribute to chemical weathering. The detail of the weathering processes that attack all exposed bedrock and contribute to the development of all landscapes will be considered in Chapter 4, but here it is necessary to look briefly at how each of the key characteristics can influence the formation of the landscape.

a) Chemistry

Rock chemistry has a significant impact on the small-scale features that are likely to be found in an area. Many features that are found at the micro- or meso-scale are the product of the interaction between the minerals in the rock and the water that falls on it or flows over and through it. Areas rich in **calcite** are particularly prone to such features, as can be seen in the swallow holes and caves of the northern Jamaican limestone or the Carboniferous limestones of north Yorkshire. In granite areas rich in potassium, feldspar (orthoclase) **kaolin** may have formed, and could now play host to major quarrying enterprises such as those at Lee Moor, near Ivybridge in Devon.

b) Resistance to erosion and weathering

Sometimes erroneously referred to as the hardness of rock, the **resistance** of any rock to erosion and weathering is a product of far more than its physical hardness. Chalk will dent if it is struck with a hammer, but its resistance to erosion can be clearly seen in the White Cliffs of Dover, and the chalk uplands of southern England. Its resistance to erosion is a consequence of its porosity; water is able to pass straight through its structure, reducing the likelihood of any erosion occurring. In direct contrast to chalk is the other significant calcareous rock. Like chalk Carboniferous limestone creates upland areas, but it is physically hard, and water is able to pass through joints and bedding planes. Granite is also a physically hard rock, such that granite landscapes often have a bleak uniformity about them, broken by the occasional tor rising above the surrounding moorland.

c) Permeability

How rocks deal with water has already been mentioned, and it is hard to disassociate this from the other characteristics. A rock's response to the attempts of water to follow gravity will contribute to the landforms that develop on that rock. Clay areas will not permit much water to infiltrate the surface, leading to higher drainage densities and greater quantities of overland flow. This leads to a generally lower and flatter landscape than in areas that have a higher permeability, as well as a greater risk of floods.

d) Structure

In geology, **structure** can be defined as the overall relationship between rock masses, together with their large-scale arrangements. Thus, the structure of rocks will contribute to the overall nature of the landscape on which the smaller scale features mentioned earlier will be superimposed. So, the boundary between the north-western edge of the chalk of the Chiltern escarpment and the clays that form the Aylesbury and Oxford lowlands are the structure of the landscape. The drainage patterns, however, are a consequence of other characteristics of those rocks, as mentioned before.

Overall, then, the landscapes found all over the world are the product of a large number of complex interactions, and the characteristics of the rocks are only a small part of that process.

2 Landforms and Landscapes of Igneous Rocks

Igneous rocks produce as many distinctive landforms as do sedimentary rocks, and they can be categorised relatively easily. The landforms produced by **intrusive** rock, those that cooled underground, are very different to those produced by **extrusive** igneous rock, those that cooled on the surface.

Extrusive landforms are usually characterised by the large distances over which the lava has flowed as it cools, creating a landscape of **basalt plateaux**, such as the Antrim plateau in Northern Ireland, which includes the Giant's Causeway.

Intrusive landforms are usually upland areas created from masses of granite that have cooled deep underground and subsequently been brought to the surface as a result of tectonic uplift (as in the case of the Sierra Nevada in the western USA) or the erosion of other rocks that had previously overlain the granite. This is the case in south-west England, where the granite that forms Dartmoor has been exposed after being formed at great depth.

a) Landforms of extrusive igneous rocks

Extrusive igneous rocks are formed by the eruption of volcanoes on to the surface. They are usually the product of lava with low silica content and therefore of low viscosity. Extrusive landforms rarely feature dramatic mountain chains or steep-sided valleys. The all-encompassing nature of a lava flow tends to ensure that the resulting landscape is rather monotonous, although it may include some features of great beauty and significance.

Flood basalts

Flood basalts are vast areas, often of thousands of square kilometres, that are covered by lava flows. It is thought that these flows were caused by long gentle eruptions of lava that built up in thickness and area because of the length of the eruption.

In the area that is now the US Pacific north-west there is one such flood basalt. Known as the Columbian basin, this area of unbroken basalt covers some 130 000 km^2, as shown in Figure 2.1. Estimates suggest that the eruptions responsible for the 2000 m thick lava flows may have lasted 15 million years, ending as recently as 12 million years ago. As many as 270 separate lava flows have been identified through the area, and they are thought to be the product of a number of

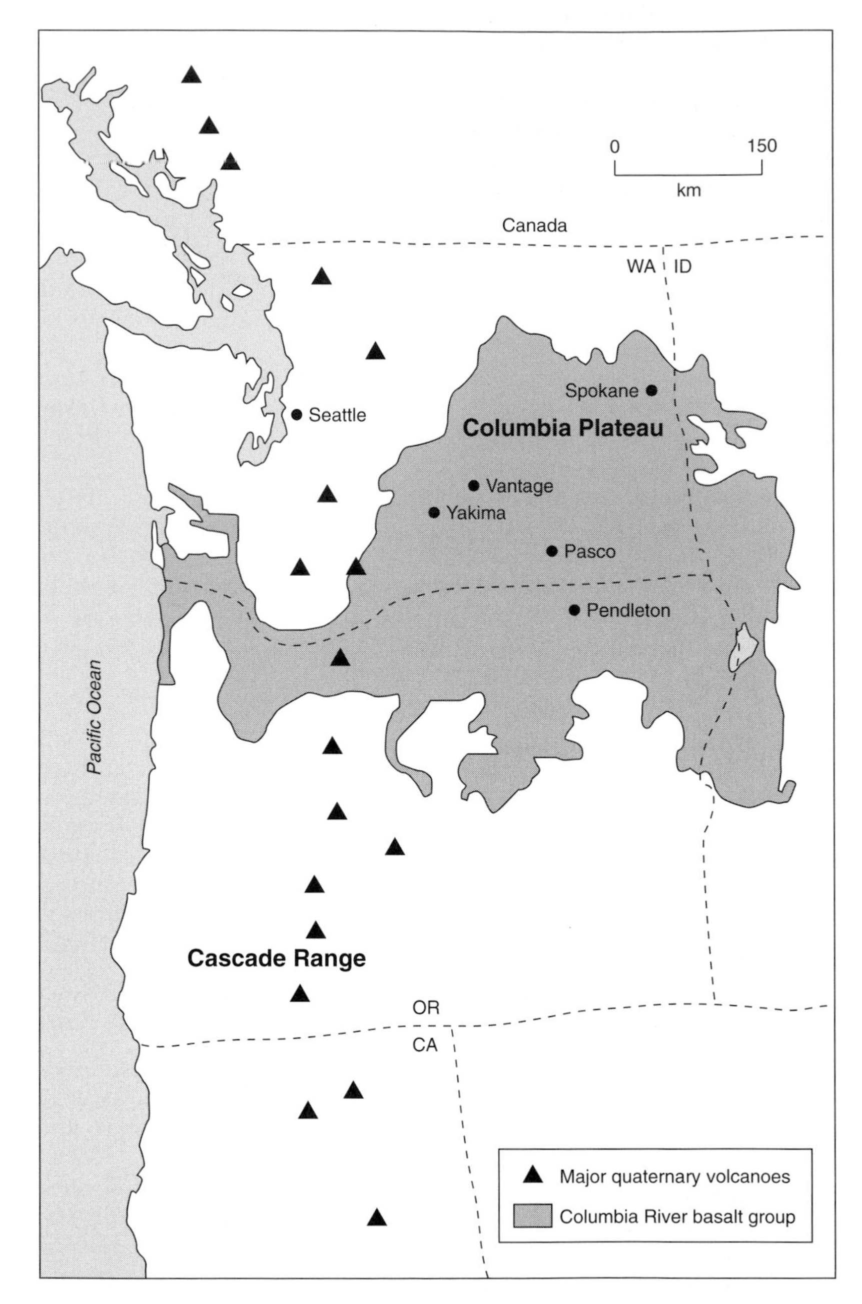

Figure 2.1 The location of the Columbian flood basalt in the Pacific north-west of the USA.

fissure volcanoes several kilometres in length, producing enough lava to cover existing relief of up to 1500 m. Although it might be expected that such an outpouring of lava would create an upland area that is not the case. The huge weight of the basalt and the space left by the erupting lava caused a gentle bowing in the crust, and the flood basalt now occupies a basin.

The Deccan plateau in western India is also thought to be a flood basalt, although one of much greater antiquity. The Deccan covers almost all of Maharashtra, and parts of Gujarat, Karnataka and Madhya Pradesh, and may well have been created as the consequence of volcanic activity. While the Deccan plateau covers an impressive 500 000 km^2 it is dwarfed by the vast Parana plateau in Uruguay and southern Brazil that covers in excess of 750 000 km^2. In both of these places, the age of the rocks suggests that the plateaux were formed by outpourings of lava associated with the break up of the ancient supercontinent, Gondwana. It has also been suggested that the eruptions that created the Deccan might have been responsible for a significant climate change that led to the extinction of the dinosaurs at the Cretaceous–Tertiary boundary some 65 million years ago.

Lava plateaux

Lava plateaux are essentially smaller versions of flood basalts. Where the scale of the flood basalts of India and the USA cover hundreds of square kilometres, a lava plateau is a smaller landscape. One of the most famous lava plateaux in the world is in County Antrim, Northern Ireland (see Figure 2.4). Others are found in Iceland. The plateaux tend to have a generally flat topography, such as can be found across much of the northern part of Northern Ireland, but at the northern coastline a remarkable exposure of basalt is visible. Equally impressive are the basalt exposures on the south coast of Iceland. At Dyrhólaey and Vik at the southernmost tip of Iceland there are cliffs made up of basalt columns similar in structure to those seen at the Giant's Causeway. As basalt cools it produces a joint pattern known as **columnar jointing**. This manifests itself in a pattern of regular polygons if viewed from above, and of tessellating columns if viewed in cross-section. As the basalt cools and contracts these cracks are formed and dictate the nature of the landscape. The joints provide a clear weakness for water to attack, and it is not uncommon for columns to collapse away from the mass of basalt when sufficient weathering has occurred.

The lava plateaux of County Antrim and Iceland had similar methods of formation. The opening of an ocean at a constructive plate margin led to outpourings of lava on a large scale. This lava flows easily across the existing landscape, filling in and smoothing out existing topography, and building up the plateau. When the Atlantic Ocean first began to open some 60 million years ago, fissures erupted lava on to what is now Northern Ireland, western Scotland and eastern Canada.

The same process continues today in Iceland as the Atlantic continues to widen and lava has flowed from fissures in Iceland.

Lava flows

Basaltic eruptions are not confined to huge fissure vents that create vast landscapes. Other volcanoes are capable of producing basaltic eruptions from time to time, even if they are interspersed with acidic, more viscous, activity. Mount Etna on Sicily has produced a wide variety of igneous material during its long and active life, and basalt flows have covered the surrounding landscape during human history. Three times between 1000 and 1400 lava from the summit reached the southern coast of Sicily, and in 1669 the port town of Catania, now Sicily's second city, was destroyed. Further back in history, a lava flow cooled to create the typical prismatic column structure on the north-east side of the mountain. Subsequent fluvial action has exploited the weaknesses provided by the joints to cut the Alcantara gorge, a spectacular steep-sided ravine with exposures of basalt on either side.

b) Landforms of intrusive igneous rocks

When magma cools underground to form granite, a wide range of features can form. Many of these features are called **intrusions**. An intrusion is a body of igneous rock that has been injected, while molten, into existing crustal rocks. After having been injected it cools without reaching the surface, thus forming igneous rocks that begin their existence underground. Some of the common forms of intrusion are shown in Figure 2.2.

Batholith

The largest of these features is a **batholith**, a large body of granite up to hundreds of cubic kilometres in size. It is thought that batholiths found reasonably close together may be joined at depth. The Sierra Nevada in the USA, the Rockies and the Andes are all underlain by batholiths of various sizes, although only the very tips have been exposed, as can be seen in the granites of Yosemite Valley in California. In south-west England, however, the batholith underlying Dartmoor, Bodmin Moor and the Scilly Isles has been exposed more fully and is being attacked by processes of weathering and erosion.

Laccolith

The massive nature of a batholith means that it pays no heed to the characteristics and structure of any existing rock. Other, smaller, intrusive features often owe some of their features to the structure of the rock into which they were intruded. A **laccolith** is a lens-shaped intrusion that has forced between the layers of the country rock and forced the overlying strata to arch upwards, forming a dome.

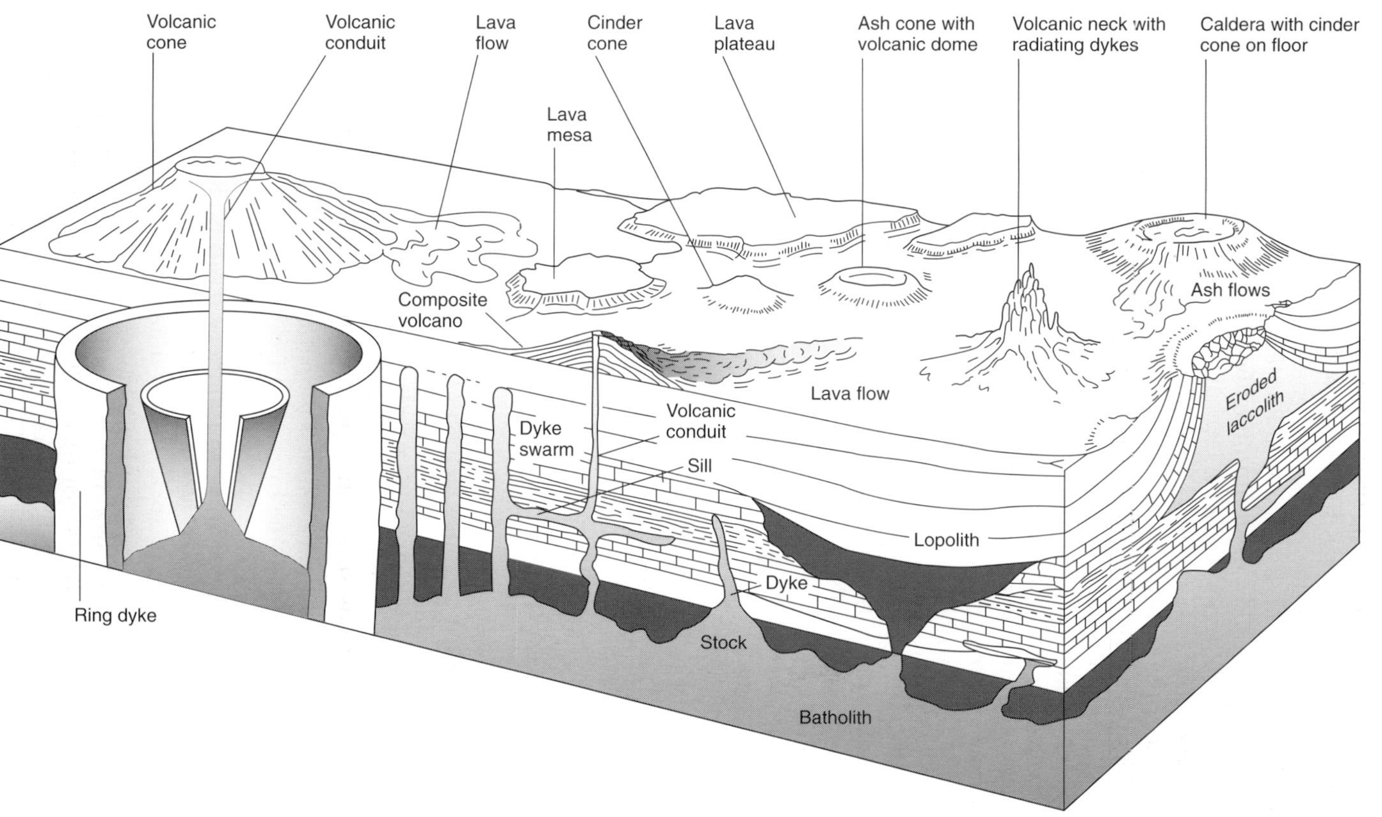

Figure 2.2 Common forms of igneous landscape. *Source*: Edwards and King (1999).

Subsequently, this dome may be weathered and eroded, leading to a small range of hills formed from an igneous rock in a landscape that otherwise consists of the rocks into which the intrusion was forced. The Eildon Hills in the Scottish Borders are thought to be the remains of a laccolith, made up of acidic lava that invaded the bedding planes of the sedimentary Devonian old red sandstone that is common in the areas around Melrose. The lavas, which are from the Carboniferous period, forced the older sandstones to dome upwards, but did not crack the strata. Had they done so, the lavas would have flowed onto the surface. The intrusive material hardened as it cooled, and remained below the surface until denudation had removed enough of the sandstone to expose the laccolith, which can now been seen as a small range of hills protruding some 200 m above the surrounding area.

Sills and dykes

While batholiths and laccoliths are perhaps best described as areas of igneous intrusion, **sills** and **dykes** are linear intrusions. As shown in Figure 2.2, a sill is an intrusion that has been injected as a near-horizontal sheet between the bedding planes of the existing rocks. A dyke is very similar except that it cuts across the existing structure to form perpendicular to the bedding planes. On being intruded, a dyke may follow the line of a joint or other weakness in the existing rock. Dykes are often found in straight lines, but sometimes they will radiate out from a central point. When this occurs the feature is known as a **ring dyke**.

Sills can be several hundred metres thick, such as the Palisades Sill in New York State, or, perhaps more commonly, a few tens of metres thick. The Great Whin Sill in northern England is between 10 and 50 m thick at different points on its exposure.

Dykes, like sills, can appear at a range of scales. They are associated with a larger igneous intrusion or with a volcanic vent, since it is from some other feature that the magma flows in order to create the dyke. At one end of the scale are the dykes associated with Ship Rock in New Mexico, shown in Figure 2.3. Ship Rock now stands some 600 m above the mainland surface, but it is thought to have been formed up to 1 km below ground. As the surrounding sandstones and shales have been weathered and eroded away, the tough volcanic rocks have resisted erosion and have assumed a prominent position. Radiating out from the rock are six dykes, two of which can be seen in Figure 2.3 as sharply defined ridges of rock. At the time that the intrusion was occurring, magma was forced sideways away from the main vent that is now Ship Rock. This intrusion would have happened at some depth below the surface and, as with the volcanic neck of Ship Rock, the dykes have been exposed at the surface because of the removal of the overlying rocks.

Figure 2.3 Ship Rock, New Mexico, USA.

Not all dykes stand up above the landscape as the ones at Ship Rock do. If the dyke is intruded into rock that is softer than that of the dyke, then differential erosion will mean that the dyke will ultimately stand above the surrounding landscape. However, if the rock of the dyke is less resistant than the rock into which it intrudes, then the dyke may form a topographic trench or gully through the landscape.

Dykes often occur in groups. It is unlikely that a volcanic event will create just one dyke, and so frequently there are numerous short lengths of intrusive rock at the surface in a small area. These collections of dykes are known as **dyke swarms**. Many of the Scottish Islands, such as Skye and Mull, have these clusters of dykes, all associated with one intrusive event deep in the past. The distribution of dykes in western Scotland is shown in Figure 2.4.

Ring dykes are a common variation on the dyke swarm, and require a survey of the general position of the dykes in a given landscape. A ring dyke is a radial arrangement of dykes, formed when pressure from the rising magma chamber has opened cracks in the overlying rock. Magma has subsequently forced its way into these cracks and then solidified, creating an arrangement where the dykes radiate out from an area that marks the centre of the upwelling magma chamber.

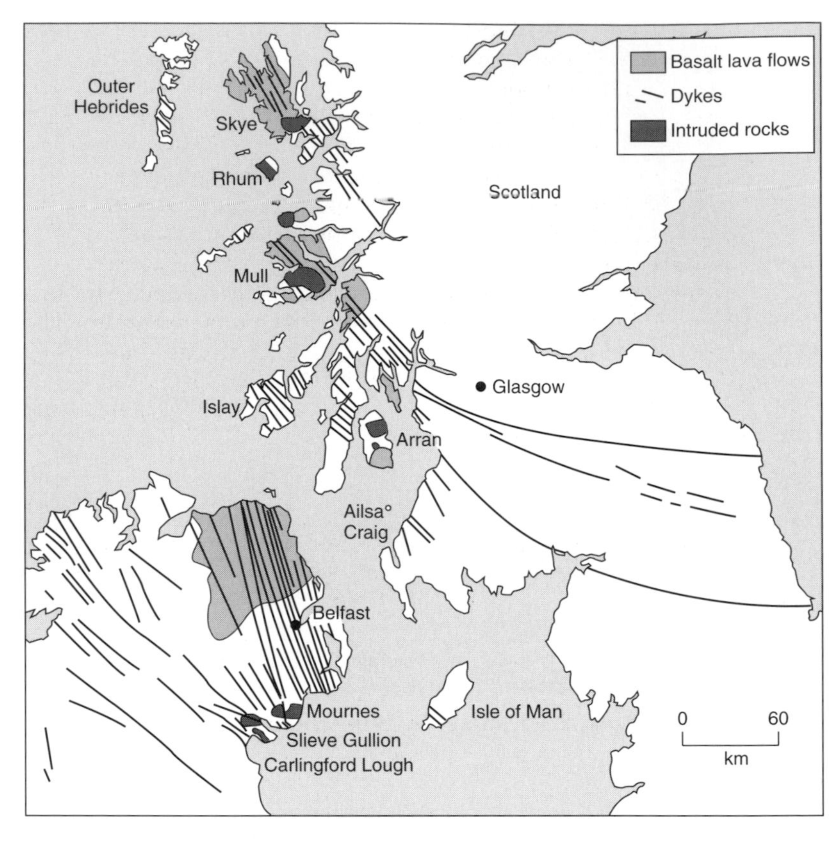

Figure 2.4 Dykes and lava flows in Scotland and Ireland.

3 Landscapes of Sedimentary Rocks

Many **sedimentary** rock types have distinctive small-scale landscape features. **Karst** topography forms on limestone, and **tors** are found in some sandstone areas. Sediment transported and deposited by the wind, called **loess**, is often home to spectacular gully formations. At a larger scale, however, sedimentary rock landscapes are usually more dependent on the input of other forces, particularly tectonic activity, to create the distinctive landscapes that have become associated with sedimentary landscapes, such as the chalk **cuesta** scenery.

a) Karst scenery

The word **karst** is derived from the Slavonic for bare, stony ground, and was first applied to scenery found in Croatia. It is characterised by an absence of or seasonal surface drainage, minimal soil cover and significant amounts of subterranean drainage. In Britain it is most commonly found on Carboniferous limestone, and is good evidence in support of the argument that geology is a crucial control over scenery. In other parts of the world karst will develop on limestone of other ages. In addition to the type sites in Croatia, karst scenery can be found in the Grands Causses of France, northern Jamaica and the Guilin area of south-west China, where spectacular **tower karst** has been formed in the subtropical conditions. The joints and bedding planes that are present in limestone are the features that exercise a distinct control over the formation of karst landscapes, since they provide a permeability that reduces surface drainage to a minimum. This absence of surface downcutting, allied to the rock's resistance, means that limestone often forms areas of high relief, such as the Three Peaks of Yorkshire, Ingleborough, Pen-y-ghent and Whernside. Several features are typical of karst scenery:

- **Limestone pavements** are formed when the blocks and grooves created by the joint structure are accentuated by the chemical action of water and exposed at the surface along one bedding plane. Limestone pavements can be found above Malham Cove in Yorkshire and in the Burren in Co. Clare, Ireland.
- **Swallow holes** form when one joint is persistently widened over time, perhaps by a surface river, to create a funnel-shaped opening leading into a subterranean drainage system. Gaping Ghyll on Ingleborough is an example of such a feature.
- Enclosed surface depressions are common in karst areas. Formed by the action of water, but not as dramatic as a swallow holes, these depressions, known as **shake holes** or **dolines**, are formed by the weathering of the limestone beneath the surface, and the associated slumping of soil into the space created. They may form the centre of a very localised drainage system. The dolines of the Yucatan peninsula, Mexico, known locally as cenotes, often contain lakes that are the focus of such a local drainage system.
- **Gorges** often form in karstic areas, usually with steep sides and large amounts of scree. Usually these are attributed to cavern collapse, although there are a number of alternative theories put forward, often because there is insufficient debris in the gorges to support this idea. It has been suggested that the Vis gorge in the Grands Causses is entirely the result of surface erosion, while Goredale Scar in Yorkshire may be the result of waterfall retreat during periglacial times. Once the water has travelled underground an equally distinctive set of features may evolve.
- **Caverns** will form when joints and bedding planes are widened as water flows along them. Eventually the openings will merge to form a cavern, such as the Carlsbad Caverns in New Mexico, USA. Debris is often found on the floor of the cavern.

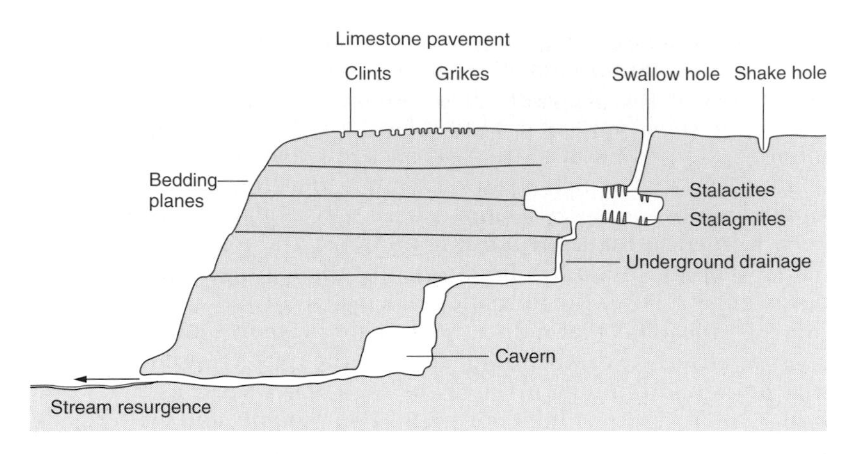

Figure 2.5 Distinctive karst features.

- **Stalactites and stalagmites** may form on the roofs or floors of the cavern. These are formed as the calcite dissolved into the water is precipitated out again when the water is stationary. If the stalactites and stalagmites grow enough to merge they will form a **pillar**. The Battlefield Cavern in Ingleborough caves, Yorkshire contains thousands of these features.

These features are common in most karst regions, and are summarised in Figure 2.5.

There are some karst features that are unique to certain conditions or environments:

- **Cockpit karst** is a very uneven landscape of depressions and low, hummocky hills. Each depression is a doline and, in the tropics, they characteristically have an irregular or star outline because as they have grown they have merged into one another. The residual hills are, in effect, the divide separating the dolines. In Jamaica the depressions are known as cockpits. The more developed scenery in this humid environment is due to the greater amount of surface drainage.
- **Tower karst** appears in humid tropical and subtropical environments such as Guilin in south-west China, and has been interpreted as a development of cockpit karst. It is characterised by steep-sided hills, up to 100 m high, divided by broad, alluvial valley floors. Here, the high levels of precipitation and increased temperatures are thought to have accelerated the processes of carbonation. The valleys contain some surface drainage, and the impact of running water has undercut the towers or **mogotes** in the landscape, enhancing the dramatic nature of these steep-sided hills.
- Tower karst has also been identified in the Mackenzie Mountains of Canada, however, and this cannot be attributed to the influence

of a subtropical climate. In this area, the limestone is massive, with few, widely spaced joints. Here it is thought that the landscape is a very well-advanced example of karstic development. The joints have been attacked for many thousands of years, and interim stages represented by dolines and swallow holes have been succeeded by a landscape of long, narrow gorges to form **labyrinth karst**, a network of deep gorges around a number of limestone towers.

Further detail on karst scenery, including examples, can be found in Chapter 4.

b) Cuesta scenery

Chalk shares few obvious characteristics with limestone, but their chemistry is essentially similar. Both are calcareous rocks, meaning that they are vulnerable to carbonation by rainwater; both allow water through their structure, although in very different ways. While the joints and bedding planes of the Carboniferous limestone provide easy routes for water to attack and eventually open up channels, the chalk forces the water to follow a tortuous route between individual particles. Thus, chalk acts rather like a sponge, and many layers of chalk have been exploited by human communities as aquifers. Chalk will form high ground because of its ability to absorb vast amounts of water. There is little significant surface drainage to lower the landscape because of the water-bearing capacity of the rock.

The main landscape characteristic of chalk is that of **scarpland**. A scarp is a slope cut into nearly horizontal strata of unequal resistance. It is characterised by a stepped profile where the exposed edge of the more resistant strata forms the steep rises and some of the plateaux, while the less resistant layers form lowlands and gentle slopes. **Simple** scarps, such as those found in the chalk scenery of southern England, have just one step, but **compound** or **complex** scarp structures may develop where the geology provides a wider variety of rock types and resistances in close proximity.

The simple scarp structure is well demonstrated by reference to the downlands of the South Downs, the North Downs and the Chilterns in southern England. During the Cretaceous era, between 65 and 142 million years ago, vast numbers of dead creatures, ranging from clams, snails and ammonites to foraminiferans (single-celled organisms still in existence today) and algae, accumulated on the sea floor. Together these made the chalk found in England today.

After its formation, the chalk was subject to shockwaves from the massive Tertiary **orogeny** (mountain-building activity) that created the Alps some 60 million years ago. This buckled to form neat horizontal patterns of sedimentary rocks and meant that much of southern England was upfolded. The amount of folding that was experienced in different places has had a subsequent impact on the nature of the landscape. On Salisbury Plain, where the chalk remained almost undisturbed, the landscape is flat plateau, while the Hogsback near Guildford is a narrow ridge formed from a band of chalk that has been turned almost through 90°, so that it is now nearly perpendicular to the ground surface. However, the most common alignment of the chalk is at a gentle angle, and this forms a **cuesta, escarpment** or **scarp and dip topography**.

The relative resistance of the chalk to weathering and erosion has led to the lowering of the impermeable clay and the creation of a steep-sided slope marking the boundary between the chalk and the clay. This is known as the scarp slope. The gentle slope running down from the summit ridge of the scarp slope is referred to as the dip slope, because it follows the dip of the layers of the rock. With distance from the summit ridge the chalk may be covered by another, more recently, deposited rock, such as London Clay. This landscape can be seen in context later in Figure 3.3.

Cuesta scenery has a number of other features:

- **Dry valleys** often form on dip slopes, and sometimes on scarp slopes. They have the patterns of normal river valleys, but are not presently occupied by a stream. It seems likely that they were formed when permanent surface drainage was present in the area. Alternatively, they may have been formed at a time when southern England was experiencing tundra climates as the Pleistocene ice sheet retreated. If this is the case, permafrost would have rendered the ground impermeable, forcing meltwater streams to flow over the surface. The size of some of the valleys in the Chilterns would seem to give credence to this theory. The Misbourne and Chess Valleys are between 50 and 60 m deep, and are occupied by streams that would never have been able to cut such a significant feature. **Winterbournes** are related to dry valleys because they are seasonal streams. During the dry months of the summer, a winterbourne will not flow, but as the water table rises in the winter a small stream will appear. The precise length of the stream will be dictated by the height to which the water table rises; the higher the water table the further up the valley the source of the stream will be located.
- Scarp slopes often house **coombes**, large hollows bitten out of the line of the scarp. Like dry valleys there are two possible explanations for their formation. The first is that springs appearing at the foot of the scarp slope would eat back into the chalk, a process known as spring sapping. The action of the water would then cause the collapse of the chalk and also hasten its removal. This theory is not supported by the large size of many of the coombes on the South Downs, or the fact that 24 of the 28 coombes between Treyford and Graffham in Sussex do not have a spring nearby at present. The

second theory suggests that the process of **gelifluction** was particularly active on such short steep slopes as chalk scarps, and that the weathered material produced by that process was subsequently removed by meltwater. Gelifluction is the downhill movement of material as freeze–thaw cycles create a saturated layer of material on the surface. It is a common process in periglacial areas.

4 Landforms of Metamorphic Rocks

Metamorphic rock does not lend itself as well to the construction of clear types of landform as either igneous or sedimentary rock. The turbulent nature of the history of metamorphic rock leads to turbulence within the rock structure and the lack of pattern is reflected in the landscape.

If the metamorphic rock has been tightly folded during its formation, its subsequent denudation may lead to the development of a landscape of ridges and valleys. More common, however, is a landscape with no particular pattern as fracture zones and shatter belts within the metamorphosed rock provide weaknesses that can be exploited by agents of weathering and erosion, while other areas close by are more resistant and provide high points in the landscape. The Green Mountains of New England in the USA are primarily ridges of gneiss, slate or quartzite intersected by valleys underlain by heavily fractured marble.

The **knock and lochan** scenery of western Assynt in Sutherland, north-west Scotland is based on Lewisian gneiss. Knock and lochan is a term coined by Linton in 1963. The terms derive from the Scots Gaelic words *cnoc*, meaning knoll, and *lochain*, meaning small lake. The landscape is full of small hills and lakes that reflect the bedrock structure. The largely impermeable nature of the metamorphic rock creates a tortuous drainage pattern linking the lakes, while the climate ensures that little grows on the sides of the hills. This landscape was created by ice moving westwards from the high land of central Scotland and taking advantage of whatever weaknesses existed in the gneiss. In places dykes have been intruded into the gneiss, and these now form low ridges, adding to the confused nature of the landscape. Human activity in western Assynt is limited to a very few **crofting** and **fishing** villages, where subsistence agriculture is a common occupation.

The Lizard peninsula in Cornwall is also a metamorphic landscape. Lizard Point is the most southerly point in the British Isles, and lies at the tip of a peninsula that extends from Cornwall. The peninsula is dominated by an open heathland plateau, cut by deep valleys. The coastline is dramatic and consists of cliffs and coves. The heath is underlain by impermeable serpentine and gabbro. Serpentine is an attractive banded metamorphic rock, while gabbro is an intrusive igneous rock. The impermeability has given rise to gently undulating land with little vegetation. Around the coastline the sea has taken advantage of the complex metamorphic geology to attack weaknesses and create coves where possible. Polurrian Cove, for example, is made of schists with clear weaknesses between the crystals. Much of the rest of the coastal landscape is cliff, however, and these have been a significant hazard for the local shipping industry during stormy conditions. Fishing villages are clustered around the sheltered bays, while the heathland area is sparsely populated because of the lack of agricultural opportunities.

CASE STUDY: DARTMOOR TORS

Dartmoor is one of England's largest upland areas. It covers some 625 km^2 of Devon in south-west England, and more than half that area is over 300 m above sea level, with the highest point at 621 m at High Willhays. The granite that forms Dartmoor was intruded during the late Carboniferous/early Permian period, some 280 million years ago. The batholith that was then formed is thought to extend from beyond the Isles of Scilly in the far south-west to Dartmoor in the east, a distance of some 180 km. The granite reaches the surface in only some of that area. Dartmoor and Bodmin Moor are the largest outcrops, while St Austell, Carnmenellis and Land's End, together with the Scilly Isles, are smaller outcrops. It is estimated that the granites of this south-western batholith cooled and solidified 6 km below the surface. It can therefore be deduced that over 6 km of the surrounding rock has been eroded away to allow Dartmoor not only to reach the surface, but also to become the highest land in the area. (See Figure 2.6.)

As the granite cooled to form the batholith, metamorphism was occurring on the edge of the intrusion to create a **metamorphic aureole**. This band of altered rocks around the batholith has proved immensely useful to humankind, since one of the main products of the metamorphism was kaolin, a valuable mineral now mined extensively on the south-west fringes of Dartmoor. Also, as the cooling took place the granite contracted and cracked, creating a series of joints and bedding planes within the rock. These joints were subsequently further enlarged by pressure release as the weight of the overlying rocks decreased. Once at the surface the joints in the granite became a prime target for weathering, allowing the formation of the landform that Dartmoor is best known for, the tor.

A tor is an outcrop of blocks of rocks. They are usually found on the summit of a hill, and they range in size from about 4 to 10 m. Their length can be much greater. Hay Tor is 70 m long and is one of the largest tors on Dartmoor. Tors are also found

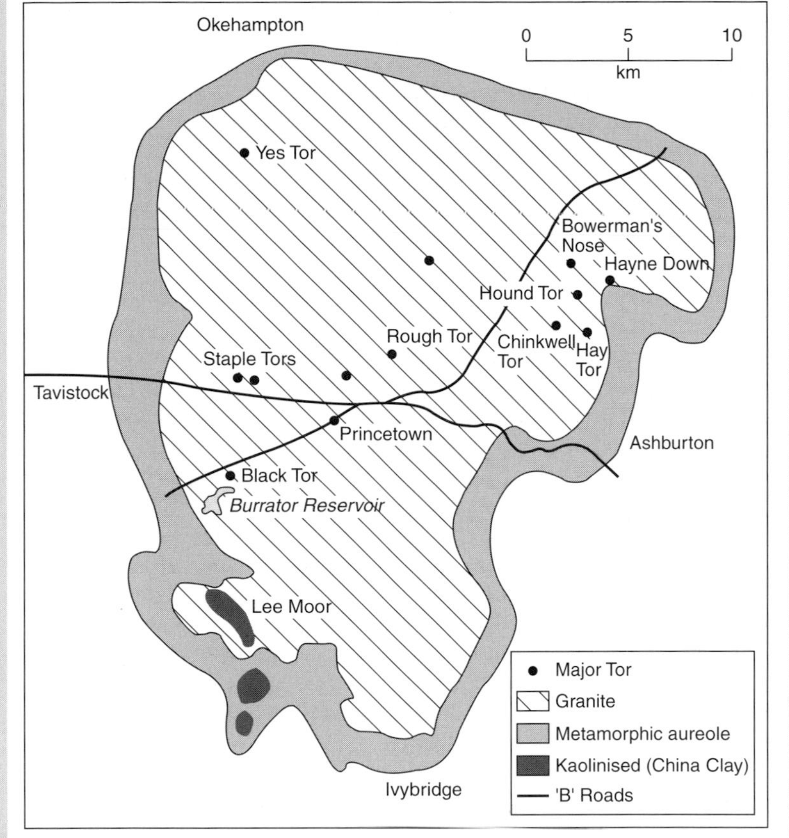

Figure 2.6 Dartmoor. *Source*: Punnett (2003).

in other locations where weathering has been focused, such as spur ends, steep valley sides and on the edge of plateaux. The boulders, or corestones, that make up tors often are heavily weathered, giving an impression of a pile of rocks loosely built up by some mysterious giant from the past. This illusion is not shattered by the bleak and inhospitable nature of the exposed and barren moorland of Dartmoor. They are surrounded by smaller blocks of weathered material that has rolled away from the tor site. These boulder-strewn slopes are known as **clitter slopes**.

Tors provide a good example of **equifinality**. This is the concept that allows for more than one plausible method of reaching the same landform. Dartmoor's tors have been exposed to weathering processes for some 10 million years and the moor has experienced many different climatic conditions during that time. This variety of climatic conditions is at the root of the problems involved in determining the formation of tors on Dartmoor. There have been at least three hypotheses put forward for tor formation:

- **Linton** (1955) suggested that subterranean chemical weathering widened the joints in between the blocks of granite in warm subtropical conditions during the Pliocene period, some 7 million years ago. The warm water would have attacked the edge of the blocks, widening the joints and forming ever-shrinking corestones. As the amount of weathered material built up, transportation processes would have removed it, eventually leaving the corestones exposed as an upstanding mass of boulders.
- **Palmer and Neilson** (1962) suggested that tors formed during much more recent periglacial times, when Dartmoor was close to the edge of the Pleistocene ice sheet and would have experienced conditions similar to present day Iceland. Intense frost shattering would have fragmented the granite, followed by rapid mass movement and solifluction to remove the broken down material and form the clitter slopes. Palmer and Neilson studied the tors carefully and noted that quartz crystals and mica flakes usually surround them. The chemical weathering proposed by Linton, they argued, would not produce such debris. However, it is thought that the Scilly Isles are too far south to have experience periglacial conditions, but granite outcrops on these islands do have tors on them.
- A third theory, that of **Bristow** (1996), combines the ideas of Linton and Palmer and Neilson, and adds a twist of his own. Bristow agrees that both chemical weathering under tropical conditions and mechanical weathering under periglacial conditions play a significant role in tor formation. Also involved, however, is a process of decay from within the granite itself. The first part of the batholith to cool is that which is nearest the surface. This forms a hard cap to the batholith, even though the interior of it is not yet solidified. This hard cap traps not only the magma, but also gases and hydrothermal liquids. These gases and fluids escaped through the joints as the granite continued to cool and weakened the granite through a process known as pneumatolysis, a form of chemical weathering caused by hot gases. This will have begun the breakdown of minerals that was continued by the subterranean waters during the Pliocene. Bristow says that this process also took place during even earlier times when Devon experienced tropical climates. Evidence to support Bristow's ideas can be found in some Dartmoor quarries where corestones are still surrounded by weathered material and there are veins of tourmaline, a mineral that is created by pneuma-

tolysis. Periglacial processes, according to Bristow, merely put the finishing touches to the long story of tor formation. Solifluction removed weathered material from the immediate surroundings of the tor, creating clitter slopes and contributed in a relatively minor way to the disintegration of the weathered granite.

Tors, then, are a complex landform. Their occurrence is not limited to Dartmoor, or indeed to granite. There are sandstone tors on Exmoor and tors on quartzite in Shropshire, and huge tors that look just like Dartmoor's except for their size in Hustai Nuruu National Park in Mongolia. However, the tors of Dartmoor are among the best known and most frequently studied.

CASE STUDY: GREAT WHIN SILL

The Great Whin Sill is one of the largest and most significant igneous intrusions in Britain. Its significance goes beyond the purely geological because it has had an influence on human activity in northern England for at least 2000 years.

The underlying rocks are a geological unit known as the Alston block. This is a unit of rock bounded by faults, made up of Lower Palaeozoic (approximately 500 million years ago) sedimentary and volcanic rock. One of the fault complexes bordering it is known as the Pennine faults. On top of these ancient rocks are Carboniferous sedimentary rocks, formed some 300 million years ago. Towards the end of the Carboniferous period tectonic activity caused igneous rock to be intruded into the weaknesses in the Carboniferous rocks and between the Carboniferous rocks and the much older rocks below. The rock that formed is a **dolerite**, which is a medium-grained rock. It has crystals that are smaller that those usually found in granites, but larger than those to be seen in a basalt. This intrusion underlies much of south and east Northumberland, as well as the Durham coalfields, but reaches its greatest thicknesses (about 70 m) further inland in the North Pennines. The name 'sill' is a quarryman's term for any more or less horizontal body of rock, and the term was first used in reference to the Great Whin Sill, and has passed into more general geological usage from there.

Scenically, the Great Whin Sill provides several interesting points. It is resistant to weathering and erosion, and therefore provides high points in the landscape. Because of its linear nature those high points tend to take the form of long lines of crags, such as those at Holwick, Cronkley and Falcon Clints, in

Upper Teesdale. These crags climb around 100 m in the form of steep scree slopes and cliffs. Resistant intrusions such as the Great Whin Sill provide a challenge to the local rivers as well. At High Force, where the River Tees crosses the sill, it is possible to see the dolerite very clearly as it forms a hard cap rock. The significance of the rock type and the intrusive feature in the formation of High Force should not be underestimated, since the waterfall is England's highest.

As with other intrusive rocks, the existing rocks next to the Great Whin Sill are subject to metamorphism. In parts of Teesdale the Carboniferous limestones have been metamorphosed into a coarse grain marble that has weathered to produce a soil capable of supporting a rare and beautiful alpine flora. Spring Gentian (*Gentiana verna*), bird's eye primrose (*Primula farinosa*), mountain avens (*Dryas octapetala*) and Teesdale violet (*Viola rupestris*) are all found on land adjoining the sill, showing the close relationship between geology and ecology that is often overlooked. The Farne Islands, off the coast of Northumberland, are one of Britain's main bird sanctuaries. The islands are made of an outcrop of the sill that has proved resistant to the erosive powers of the sea, providing an offshore haven for birdlife.

Human uses of the features of the Great Whin Sill are also varied. There is evidence of quarrying of the dolerite going back well before the Roman occupation, and quarrying continues to this day. The dolerite of the Great Whin Sill is resistant to impact and attrition, and is therefore very suitable for roadstone use. Although quarrying on the Great Whin Sill is no longer a particularly significant industry, during the interwar period large quantities of the dolerite were extracted and used to build roads.

However, the most famous use of the Great Whin Sill was by the Emperor Hadrian. During the Roman occupation the Picts caused a great deal of trouble, and in AD122 Hadrian ordered the building of a defensive wall along the line of the Great Whin Sill. The raised ground provided by the Sill, albeit only a few metres in places, provided enough of a natural vantage point for the wall to enhance and thus provide a fine view of the land to both the north and south. Hadrian's choice of site had another advantage, since the stone that could be extracted from the dolerite made perfect building materials. The resilience of the stone is clear given that the wall stood for several centuries after the Romans left the area, and plentiful evidence is still available at locations such as Housesteads and Vindolanda. The length of the sill assisted Hadrian as well. His wall stretched from Wallsend to the Solway Firth, a distance of over 100 km, and along much of that distance it followed the line of the sill.

CASE STUDY: IGNEOUS LANDFORMS ON ST HELENA

On the map St Helena appears as a pinprick in the Atlantic Ocean, 16° south of the equator and 5° west of the Greenwich meridian. It is a volcanic island sitting on a transform fault of the mid-Atlantic ridge that can only be reached by sea. The nearest land is Angola, some 1950 km to the east. The island is 16 km long and 10 km wide, and represents the summit of the volcano that emerges from the seafloor 4224 m below. The base of this volcano is 130 km in diameter, and the volume of the cone is estimated to be 20 times that of Mount Etna, Europe's largest volcano.

The summit of the volcano is 823 m above sea level at its highest, and is in the form of a long ridge running along the long axis of the island from north-east to south-west. Running off this ridge are a series of valleys cutting deeply through the volcanic material. The coastline is also rugged; 300 m high cliffs dominate the coastline at various points, and these, together with the vagaries of the Atlantic Ocean, make St Helena an inhospitable place to live. Indeed, it is in places such as St Helena that an understanding of the geology and physical geography can make a great difference to the quality of life. The nature of the terrain influences everything, from communications and settlement to agriculture and economy.

The island was formed by the **coalescence**, or joining together, of two shield volcanoes. These volcanoes have not been active to human knowledge, but have clearly erupted many times with great vigour in order to produce such a significant landform. In the north-east volcanic activity was centred on the Flagstaff Hill area. This volcano was active above present sea level from around 14.6 million years ago, and most of the basalts produced on the island date from between 10 and 13 million years ago. The second centre of activity was in the south-west around Sandy Bay and this did not become active above sea level until about 11.3 million year ago. The Sandy Bay fissure is thought to have produced over two-thirds of the material that makes up the island of St Helena as it stands above the waves today.

The valleys running from the central ridge cut through the layers of volcanic debris that have been produced. Exposures in the valley sides show lava flows interspersed with ash and pyroclastic deposits, with basaltic lava flows by far the most common. It is thought that present-day activity on Iceland and in the Hawaiian islands is a good model for activity on St Helena during the period that the volcano was active. In addition to basaltic lava flows, the island has a number of intrusive features as well. Plugs, domes and dykes can all be found and they are made of more viscous rocks called trachydolerites and phonolites.

Figure 2.7 shows a simplified version of the igneous geology of St Helena.

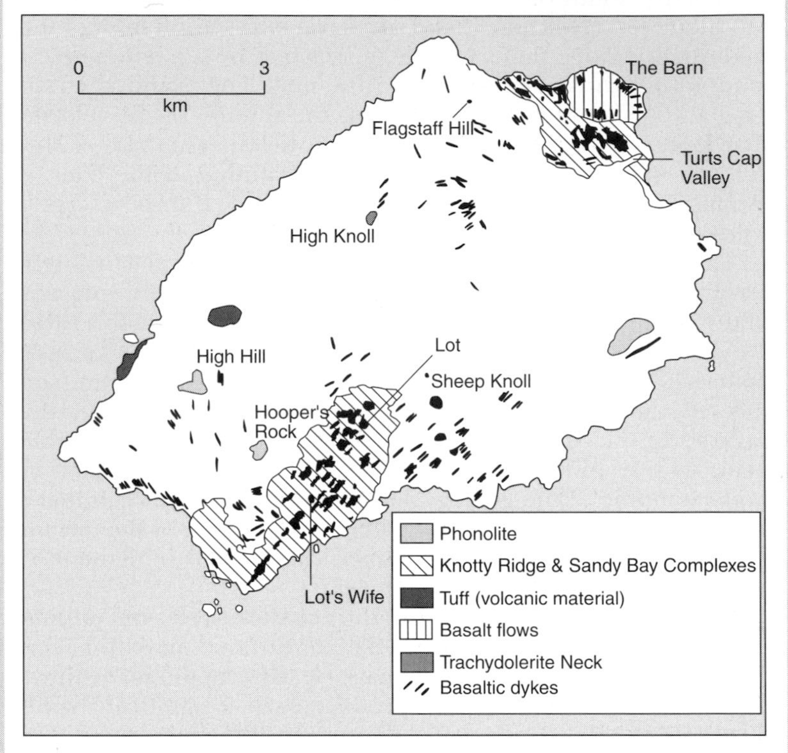

Figure 2.7 Simplified igneous geology of St Helena.

The north-eastern part of the island contains some of the most dramatic and inhospitable scenery on the whole island, and has extrusive and intrusive features in close proximity. Flagstaff Hill is one of the highest points on the island, and marks part of the summit ridge. From Flagstaff Hill a basaltic lava flow extended in a north-easterly direction to form a cap on The Barn, a 300 m high hill that has been protected from erosion by the presence of the basalt flow. Turk's Cap Valley, on the south side of The Barn, is among the most heavily dissected layers of volcanic ash and pyroclastic material on the island. This area is also populated by a vast dyke swarm of some 200 dykes. The presence of so many dykes suggests that they were formed at a time of great tension in the cone. This tension would have stretched and weakened the existing landscape, providing weaknesses for the igneous material to surge into before cooling.

In the southern part of the island, produced by the Sandy Bay fissure, different processes have been at work, but the results are no less impressive. Again, weak layers of pyroclastic material have been removed easily, leaving dykes to provide a pattern of sharp ridges. However, there are also several intrusions of the more resistant phonolite in this area. These have been isolated to form upstanding volcanic necks such as Lot and Lot's wife. These significant intrusions are thought to represent the last stages of volcanic activity, some 8 million years ago, and probably did not disturb the surface when they were created. It has been estimated that the intrusion of these phonolites was the end of some 40 million years of volcanic activity, of which the last 6 million or so saw the construction of that part of the volcano currently above sea level. Lot and Lot's Wife are not situated on the same magma source. Although they both formed as swellings on a dyke, they are on different dykes radiating from the volcanic centre.

In many places in the world igneous activity has provided a rich source of income for the communities that inhabit the area, although this cannot be said of St Helena. As a UK dependency, the island economy is reliant on British aid, and it was hoped in the 1950s that mineral resources would provide a means of independence. Manganese and phosphate were found in the island, but in insufficient quantities to be economically viable, especially given the huge costs involved in transporting anything to or from St Helena. Some hard core is quarried from basalt to maintain the 80 km of surfaced roads on the island, but the main economic activity on the island is growing and exporting coffee, which has been cultivated on the steep slopes of the valleys since the 1730s.

Summary Diagram

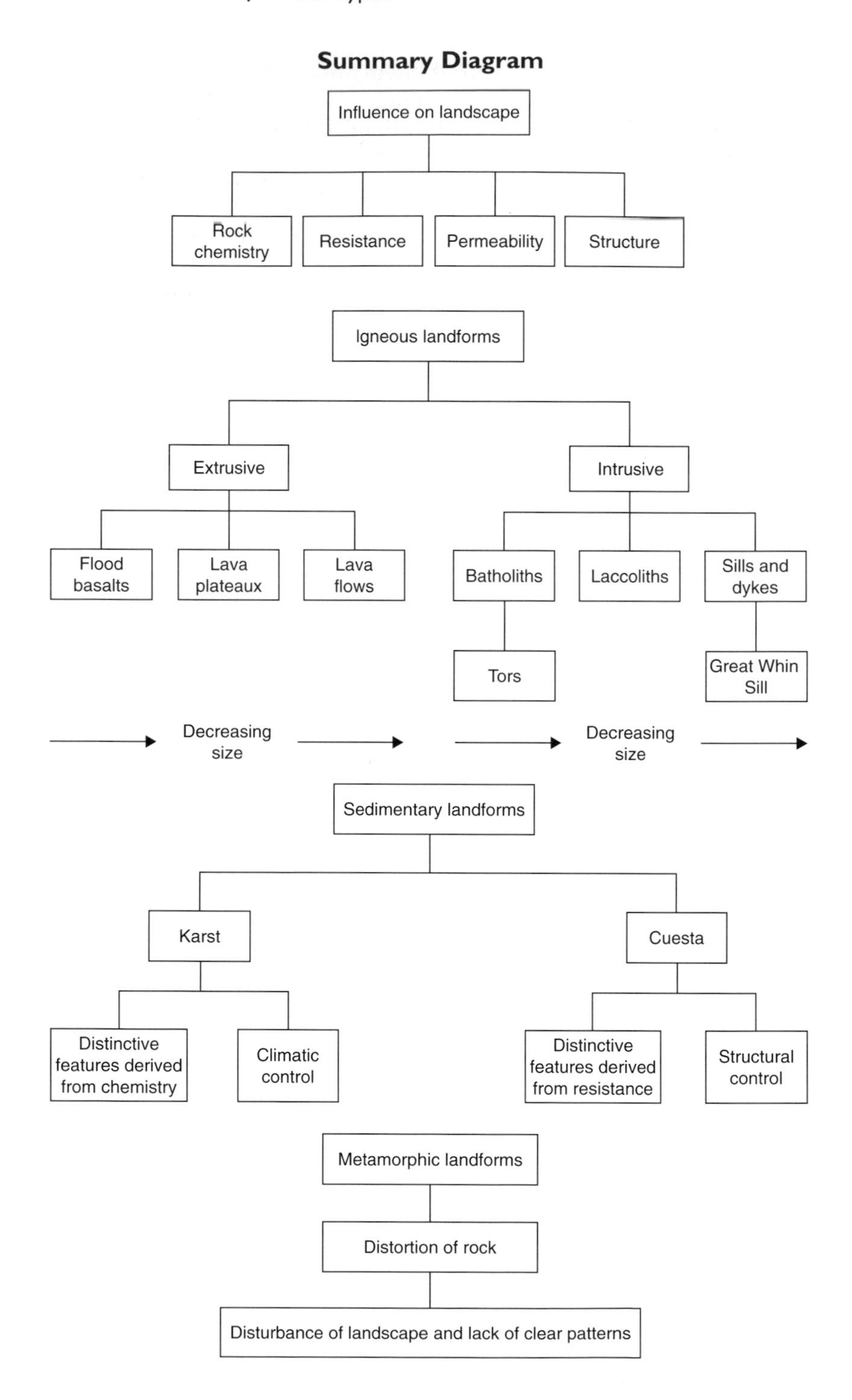

Questions

1. Show how rock type may dominate landforms in some areas. Use examples of different rock types to support your arguments.
2. State the differences between the various proposed theories of tor formation.
3. Draw a geological sketch map of the area shown in Figure 2.3. Annotate it to explain the landscape created.
4. **a)** Why are caves common in areas of Carboniferous limestone?
   b) Explain how stalactites and stalagmites are formed.
5. Why does karst landscape develop differently in various parts of the world?
6. Explain the formation of dry valleys and coombes in chalk landscapes.

3 Landscape of Geological Structures

KEY WORDS

Scale: the size or magnitude of a feature or event. Geological structures occur at a huge range of scales.
Structure: the relationship between rock masses, along with their large-scale arrangement.
Anticline: an arched upfold in the layers of the Earth's crust.
Syncline: a downfold or basin formed from layers of the Earth's crust.
Tension: forces applied to layers of the Earth's crust that stretch them.
Compression: forces applied to the layers of the Earth's crust that squash or squeeze them.
Fault: a rupture or crack in rock layers caused by the application of force, and where displacement of the layers is visible.
Fold: a bending of the layers in the Earth's crust caused by compressional forces.

1 Scale

Although different rock types can develop distinctive landscapes, as demonstrated in the previous chapter, this is not the only part of geology that can have an influence on the present-day landscape. The rocks that we see at the surface today, and that contribute to the landscape of that surface, have been subject to a considerable number of forces before reaching their current location and state. Many of those forces are movements that are created by tectonic forces. The causes and nature of tectonic movements does not form part of this book, but the landscapes that have been created by those forces do because they are part of the picture of landscapes on the surface of the Earth.

Landscapes created at geological structures can be considered at a number of **scales**. Geomorphology is concerned with phenomena on the surface at an enormous range of scales, from an individual boulder or rock to the size and shape of major relief features such as mountain ranges. In order to do this geomorphologists have adopted a category for features of different sizes, although, as with any geographical categorisation, it can be rather hard to decide where to draw the boundaries.

- **Mega-scale** landforms are the largest sorts of landforms. These include major mountain ranges such as the Himalaya and equally

substantial drainage basins like the Amazon. Processes that influence such features work over millions of years and include tectonic processes and very long-term climate change.
- **Macro-scale** features include parts of mega-scale features, perhaps the currently glaciated area of the Himalaya or the floodplain section of major drainage basins. Processes that act on such features will take many years to have a major impact.
- **Meso-scale** landforms are individual landforms such as a volcano or a meander within a river. Processes that have an impact on these landforms will be ones that are familiar to most students of geography, such as river or glacier flow and localised volcanic activity.
- **Micro-scale** features are those that occur within a meso-scale feature, such as riffles and pools in a meander, or dykes and sills within a volcano. These features are created by processes that function over short spans of time, even as short as a few days in the case of an individual flood event.

The largest and smallest of these categories are probably the least well understood, while most study is directed at the macro- and meso-scale features.

Geological processes function at all of these scales. Of all the other significant groups of geographical processes perhaps only **climate** can be said to do the same. Geomorphological processes such as glacial or fluvial processes function to create landforms at the three smaller scales, but they only work on landscapes of mega-scale, because the largest features have been introduced into the landscape by another set of processes. Major **mountain ranges**, for instance, have been created by geological or tectonic processes, while features like **deserts** and **rainforests** are largely the product of climatic influences.

Human activity cannot hope to influence the mega-scale features, except perhaps through long-term climate change, although features at all other scales are vulnerable to alteration from human activity, for example through the construction of dams such as the Three Gorges dam in China.

2 Geological Landforms at Mega- and Macro-scales

The largest categories of landforms are often created from geological structures that have in turn been created by the global processes of plate tectonics. The mega-scale features that result from **geological structures** are those formed at plate margins, such as **island arcs** and **fold mountains** produced at continental collision zones. Examples of these two features are the Japanese island arc produced at the boundaries between the Pacific, Eurasian and Philippine plates or the Himalaya, formed from the collision of the Eurasian and Indian plates.

Macro-scale features that arise from tectonic activity are present along part of a given plate boundary rather than being thc product of its whole length. Examples of macro-scale features include **rift valleys** forming at constructive plate boundaries in Iceland and East Africa and areas of significant **folding** such as the Zagros mountains in Iran.

a) Rift valleys

Rift valleys are important landforms created where tensional stresses dominate the crust:

- They typically have widths of between 30 and 60 km, although some parts of the East African rift are as much as 100 km across, and may run for several thousand kilometres across the landscape.
- They are broad, flat-floored and usually symmetrical troughs, flanked by steep-sided escarpments.
- They represent the splitting apart of the Earth's crust, and may occur on an existing constructive plate margin, such as at Þingvellir in Iceland, or the development of a new margin, as is found along the East African rift.
- Each side of the valley marks the location of a **fault**, along which the central part of the rift valley has dropped, as shown in Figure 3.1. The height difference from the top of the ridges to the valley floor can be as much as 2000 m. The area between the two faults is known as **graben**.

Rift valleys such as those in Iceland and East Africa are simple to comprehend because the action of the tension within the crust is clear.

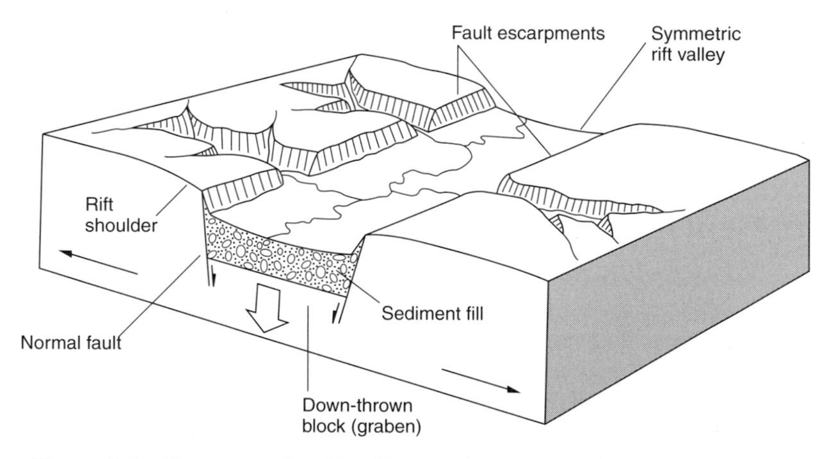

Figure 3.1 Structure of a rift valley. *Source*: Summerfield (1991). *Global Geomorphology: An Introduction to the Study of Landforms*, 1991, Pearson Education Limited

The East African system is 3000 km long, and extends from the Red Sea to Malawi in the south. As the tension is applied it creates stresses within the crust that force the rock to stretch and then crack, forming faults. Once these cracks have appeared, as shown in Figure 3.1, the area of land between the faults is forced to drop. In much of the East African system the height difference between the top and bottom of the fault is between 400 and 2000 m. The area in between the faults is known as the **down-thrown block**, and forms the bed of the rift valley. Although the feature as a whole is a macro-scale landform, the individual faults that occur within the valley to create the down-thrown block are much smaller-scale features, demonstrating the difficulty in treating any landform in isolation.

The floor of the valley is occupied by a number of lakes such as Lakes Tanganyika, Turkana and Nyasa. In addition to the lakes, a vast quantity of sediment has been deposited in the valley floor as the steep sides of the valley are eroded and weathered. Lake Tanganyika is 1400 m deep, and its floor is some 650 m below sea level, demonstrating the scale of the valley. Many of the other lakes do not form the source for a river, rather they act as the ultimate destination for streams that flow into them. As a consequence they are salt lakes because of the continuous process of evaporation that occurs without any removal of the salts created. Lake Victoria, on the other hand, is perched on a high plateau, or **horst**, that sits between two arms of the valley as it splits in its southern section.

Perhaps inevitably, given the weaknesses in the crust that must exist in an area under such tensional stresses, several major volcanoes have evolved. Mount Kilimanjaro and Mount Kenya have been built close to the valley and now reach about 6000 m.

b) Fold mountains

The term **fold mountains** is often applied to whole ranges of mountains created by the crustal thickening and distortion associated with collision zones and destructive plate boundaries such as the Himalaya and the Andes. Other areas of the world are also covered by mountains or hills that have been created by the **folding**, rather than faulting of the crust. Folding will occur when the crust is subject to compression rather than tension. This compression forces the rocks to buckle and squash. The amount of pressure applied to the surface will dictate the amount of folding that happens and the extent of that folding. Folded terrain at a macro-scale has a number of typical features:

- Parallel areas of high ground, known as **anticlines**, representing the areas that have been forced upwards by the folding.
- In between the areas of high ground there will be low land, taking the form of valleys. These are **synclines** and are the rocks that have

been folded downwards as a result of the compression. Figure 3.2(a) shows the relationship between anticlines and synclines.
- Linear drainage patterns, as rivers follow the synclinal valleys in between the anticlines; occasional breaches of an anticline may produce a spectacular gorge.

3 Landforms of Meso- and Micro-scale Geological Structures

Many of the landscapes that occur at mega- and macro-scales and that have been created by geological processes contain landforms that are of a much smaller scale, but that can also be classified as landforms of geological structures. Plate margins are invariably regions of active **faulting** and **folding** at a variety of scales, and landforms often develop that relate to crustal movements along individual faults. These landforms can occur within the product of much larger scale tectonic activity, as the case study of the Himalaya demonstrates.

a) Faulting

At the meso-scale faulting takes place as the result of sudden rapid movements within the crust ranging from a few centimetres to several metres. Each individual movement is associated with an earthquake. At larger scales, faulting is interpreted as movements along a substantial length of a plate boundary, hence the use of descriptions such as the San Andreas fault.

Faulting can occur in many ways, but the two most common are **normal** and **reverse** faulting. These terms illustrate the relative movement between the crust on either side of the fault line. In a normal fault the one block slips down relative to the other and the **fault plane** (the line of the fault, in three dimensions) forms a slope that is similar to that of a scarp. In a reverse fault one block is forced over the other, creating a fault plane that is overhung. Normal faults are usually formed by tensional action, while reverse faults are more commonly created by compressional forces. The landscape created by faulting is likely to be affected by the scale and antiquity of the faulting.

Fault scarps can be created by any faulting that reaches the surface. The initial angle of the scarp will be dictated by the angle of the fault, and whether it is a normal or reverse fault. However, over time it is inevitable that processes of weathering and erosion will destroy the original form and create a slope that is stable, typically between 20° and 40°.

Normal faults are not usually isolated features. They often occur in groups, commonly as a set of parallel faults. When this occurs a pattern will develop in the landscape to reflect the structure imposed by the faulting. A narrow block dropped down between two normal faults is known as a **graben**, while a narrow block forced up between two normal faults is referred to as a **horst**. One example of a horst block is the Ruwenzori Mountains in the western part of the East African Rift Valley. Where the rift valley splits in two on the border between Uganda and the Democratic Republic of Congo a block has been forced upwards to form mountains that reach nearly 4000 m above sea level.

Faults also provide lines of **weakness** in the landscape that forces of weathering and erosion can exploit. The importance of joints and bedding planes within rock has already been discussed in this context, and faults are, in many ways, no different. Fault-guided river valleys are common, and in many upland areas these have been turned into fault-guided glacial troughs such as Moffatdale in southern Scotland. Coastal erosion is also strongly influenced by faulting, and coastlines such as the southern shore of the Gower peninsula in South Wales has many inlets that have formed as the marine processes have exploited fault lines running perpendicular to the coast.

b) Folding

The influence of **folding** on the landscape is similar to that of faulting. At a macro-scale folding will influence the entire landscape, as in the example of the Zagros mountains in Iran. However, not all folded landscapes occur at that scale. All folding occurs as the result of **compressional forces** being applied to the landscape. The severity of the fold will be determined by the amount of force applied and the resistance of the rock to that pressure. There are several types of common fold, shown in Figure 3.2.

- The **simple fold** will form as pressure is first brought to bear on the rock structure. The rock begins to buckle, but maintains a smooth curve.

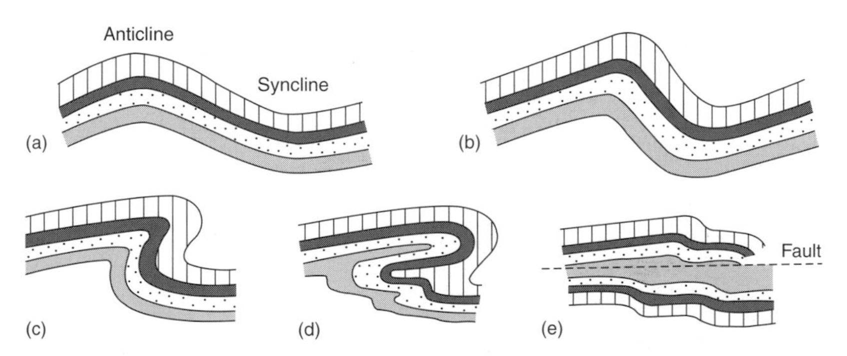

Figure 3.2 Types of fold. (a) Simple, (b) asymmetric, (c) over, (d) recumbent and (e) nappe structure. *Source*: Collard (1988).

- Further pressure develops an **asymmetric fold**, where one of the slopes dips more steeply than the other.
- Further pressure will create an **overfold** where the beds in one limb of the fold have been rotated so far that they are now more that 90° removed from their original alignment.
- The next stage is a **recumbent fold**, where the rotation of one limb has reached nearly 180°.
- This is followed by the final stage, a **nappe** structure, where the pressure has provided more pressure than the rock could bear and part of the fold has detached itself from the other part, forming a fault through the centre of what was previously a fold, and a very confused geological pattern.

Many mountain ranges owe much of their construction to the creation of folds, but at a very large scale. At the micro-scale folds can often be seen within cliff structures, especially of sedimentary rocks, while meso-scale features can be seen within the major mountain ranges. The Alps and the Himalaya are full of examples, such as the recumbent fold in the young sedimentary rocks making up the Rotstock peak in Switzerland. Equally, the scarp and dip topography of the Chilterns, described in the previous chapter, owes a good deal to the asymmetrical fold that was created as a result of the Tertiary orogeny. The gentle angle of some of the beds creates the dip slope running down from north-west to south-east towards London, as can be seen in Figure 3.3.

Inverted relief is a development of folding once the folding has occurred and the landscape is subject to processes of denudation. As the rock that is forming an anticline arches upwards and folds it will experience some stretching and possibly cracking. Conversely, the rock that is being squeezed to form a syncline will be squashed more tightly together as it is forced downwards. Thus, the anticline will have had weaknesses placed within it by the compressional forces and these weaknesses could be more easily attacked by forces of weathering and erosion. Add the fact that anticlines are forced upwards, into areas where frost shattering is more likely and where river and glacier movement will have more gravitational energy, it is understandable that denudation is likely to occur on anticlines. Synclines, on the other hand, will receive the debris produced on the anticline as deposits, and the rock that was strengthened by compression may well also be covered up and protected. As with asymmetrical folding, the south of England provides a good example of inverted relief. The Weald of Kent, shown in Figure 3.3, was once an anticline of the same asymmetrical fold structure that includes the Chilterns. It has subsequently been denuded by fluvial and, probably, periglacial processes, and is now an area of lower land between the North and South Downs.

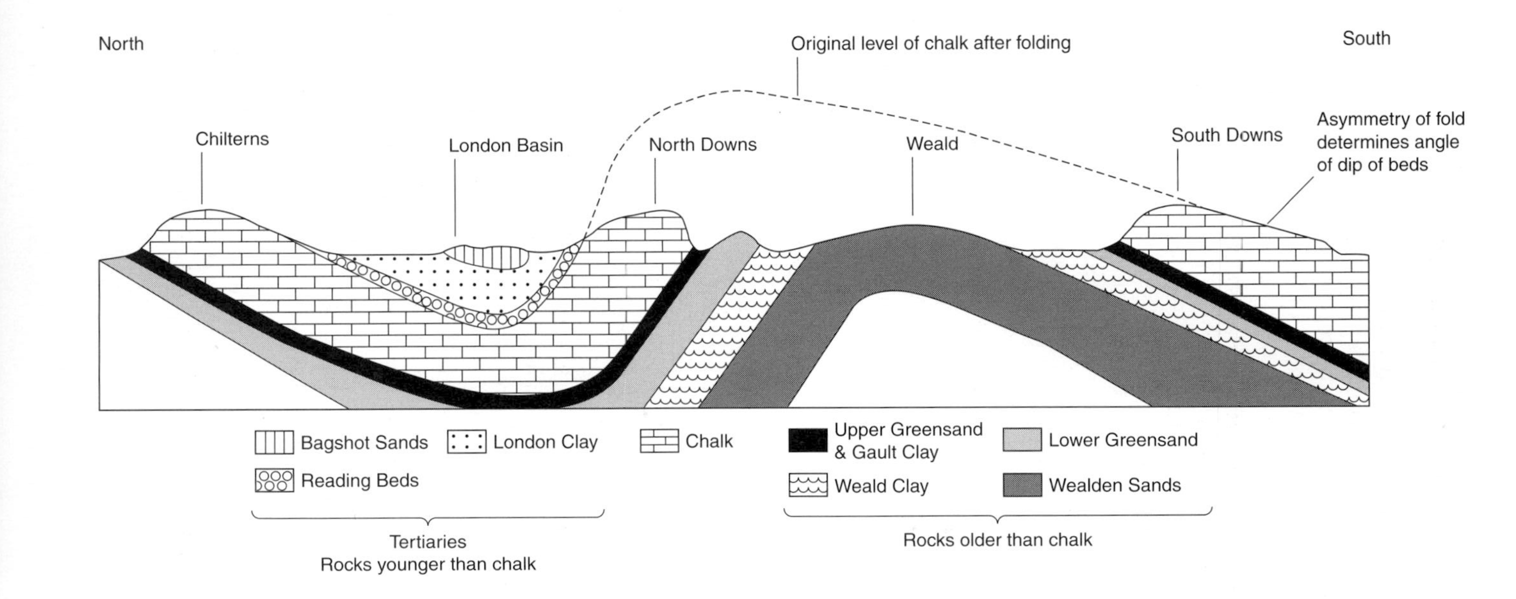

Figure 3.3 Inverted relief and cuestas in south-east England. *Source*: Collard (1988).

CASE STUDY: THE HIMALAYA

The Himalaya are currently the largest range of mountains on the surface of the Earth. There may well have been other ranges in past geological times that exceeded the present dimensions of the Himalaya. It is thought that the ancient mountain chain of the Caledonides and the Appalachians were once larger, although the remnants are now split in two by the Atlantic Ocean. The Appalachians are in north-eastern USA, while the Caledonian mountains have their remains in the British Isles and Canada. The Himalaya stretch for some 2500 km from the Indus river in the west to the Brahmaputra in the east. India, Pakistan, Nepal, Bhutan and Tibet all include part of the range, which is bordered on the north by the valleys of the upper stages Indus and Brahmaputra. These valleys are structurally controlled, since they flow along a fault line that has provided a weakness for the rivers to exploit. To the south, the mountains are bordered by a sediment-filled basin known as the Indo-Gangetic plain. This is a low-lying and generally flat area that has been filled with sediment that has washed down from the Himalaya as they have risen upwards.

A study of the Himalaya can provide evidence for many of the ideas that have been covered in this chapter, as well as some from other areas of geography.

The Himalaya range is still being formed today, but the collision between the Indian subcontinent and the Eurasian plate began some 50 million years ago, during the Oligocene period. Significant amounts of uplift did not occur until about 35 million years ago; however, the intervening 15 million years was occupied with the crumpling and deforming of sediments, and the subduction of oceanic material. Before that time the Indian plate had been travelling towards the Eurasian plate at rates in excess of 10 cm per year, a rapid rate of movement for a continental plate. It pushed ahead of it a number of continental fragments or **microplates** that were successively added to the Eurasian plate to make up what is now the Tibetan plateau. As a consequence, the Tibetan plateau is not a uniform continental interior made up of the same rock, but a collection of miniature plates joined together by **sutures**. These sutures are similar to faults in that they provide a line of weakness that can be attacked by forces of weathering and erosion. They are often identifiable because the rocks on either side of the suture are very different. For example, on the northern side of Nanga Parbat in Pakistan, reddish rocks from the Indian plate have been sutured to greenish rocks that were once part of the Eurasian plate. The main zone of sutures in the Himalaya is to the north and north-east of the main axis of the mountain chain, on the Tibetan plateau.

The history of the Tibetan plateau is one of many collisions, although the latest, that with the Indian plate, is by far the most dramatic.

The Indian plate continues to force its way northwards at about 5 cm per year or about twice the rate of growth of a human fingernail. Given the resistance that is offered to the movement of the Indian plate from the Eurasian plate this is a remarkably rapid rate. This movement causes the rocks around the leading edge of the Indian plate to deform and fracture. Similar forces are applied to the rocks on the southern edge of the Eurasian plate, and they too will buckle and crack. Vast slices of the Indian crust were pushed southward as the Indian plate forced its way north underneath the surface rocks. These slices were stacked up to form the Himalaya, and are separated from one another, and from the rocks of the Indian plate, by a sequence of large faults that dip right through the mountain range from south to north, as illustrated in Figure 3.4.

As the continents came together the rock must be squeezed, forcing material to be displaced. This occurs as material is uplifted and as a result of faults forming. Once a fault has formed rock layers can slide over each other. These processes combine to cause **crustal shortening**, and it is estimated that 400 km of crustal shortening has taken place in the Himalaya.

Marked on Figure 3.4 are some of the faults and packages of rocks that make up the Himalaya. The Indo-Gangetic plain is on the far left, and is made of rocks that are still firmly associated with the Indian plate. The 'main boundary thrust' marks the join between the Indo-Gangetic plain and the rocks of the lesser Himalayan sequence. This collection of rocks is made of ancient metamorphosed rocks that have been forced upwards as the orogeny has progressed. The lesser Himalayan sequence is separated from the greater Himalayan sequence by the main central

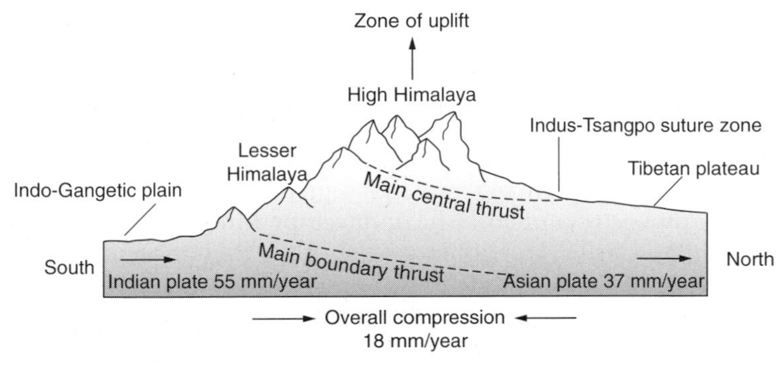

Figure 3.4 Cross-section of the Himalayan mountains.

thrust, another series of faults. The greater Himalayan sequence contains some very ancient metamorphic rocks that have been forced upwards, but also a good deal of granite and other igneous intrusions that date from the Tertiary period, and would have been intruded into the existing rocks as oceanic material was subducted under the Eurasian plate. After intruding the granite into the Eurasian plate the continuing movement of the Indian plate forced those granites to rise high above sea level. The fact that these intruded rocks are now found at great altitude has been a key piece of evidence as scientists have pieced together the story of the Himalaya.

In places, such as Kashmir and the Everest region in the greater Himalayan sequence, the highest rocks of all are sedimentary rocks that formed in the shallow conditions of the continental shelf that would have existed between the two plates as they came together. Indeed, rocks collected from the summit of Everest are relatively unmetamorphosed limestones, containing fragments of coral.

Many ranges of fold mountains have formed when vast quantities of sedimentary rocks that collected in basins were distorted and uplifted. This is not true of the Himalaya; they have been formed by the peeling off and forcing upwards of rocks that were part of the Indian continent. Fundamental to this process is an array of faults that run the length of the mountain range and provide the weaknesses along which that peeling can take place.

The Himalaya area is unquestionably a mega-scale feature, created by tectonic processes. However within that huge feature a very important part is played by the faults, which are much smaller scale phenomena.

In addition to the faults many overfolds and nappe structures are also identifiable, including one within Nuptse, a peak of 7861 m just to the south-west of Everest. Within the main body of the Himalaya faulting is by far the commonest geological structure because of the extreme forces involved and the rapidity of their application. However, on the boundary between the Indo-Gangetic plain and the lesser Himalayan sequence, the forces are being applied to landscapes that have not yet been significantly distorted. The Jhelum river, a tributary of the Indus, flows from Mangla lake in Kashmir, north-eastern Pakistan. The lake is in an area that is experiencing the first signs of mountain building; and a number of anticlines have developed in the area over the last 3 million years. The upper course of the Jhelum wends its way around these anticlines, such as the Mangla Samwal anticline, in order to find a route to join the Indus, as shown in Figure 3.5. The number of millions of years ago (mya) that each anticline formed is shown.

The locating and studying of active faults is an important geological job. It helps scientists understand the processes and patterns behind the formation of mountain ranges like the Himalaya, but it also has a significant human impact because it is active faults, however large or small, that can cause earthquakes. Studies in Bhutan, a tiny Himalayan kingdom, have shown that there are many small faults to accompany the major Himalayan thrust zones already discussed.

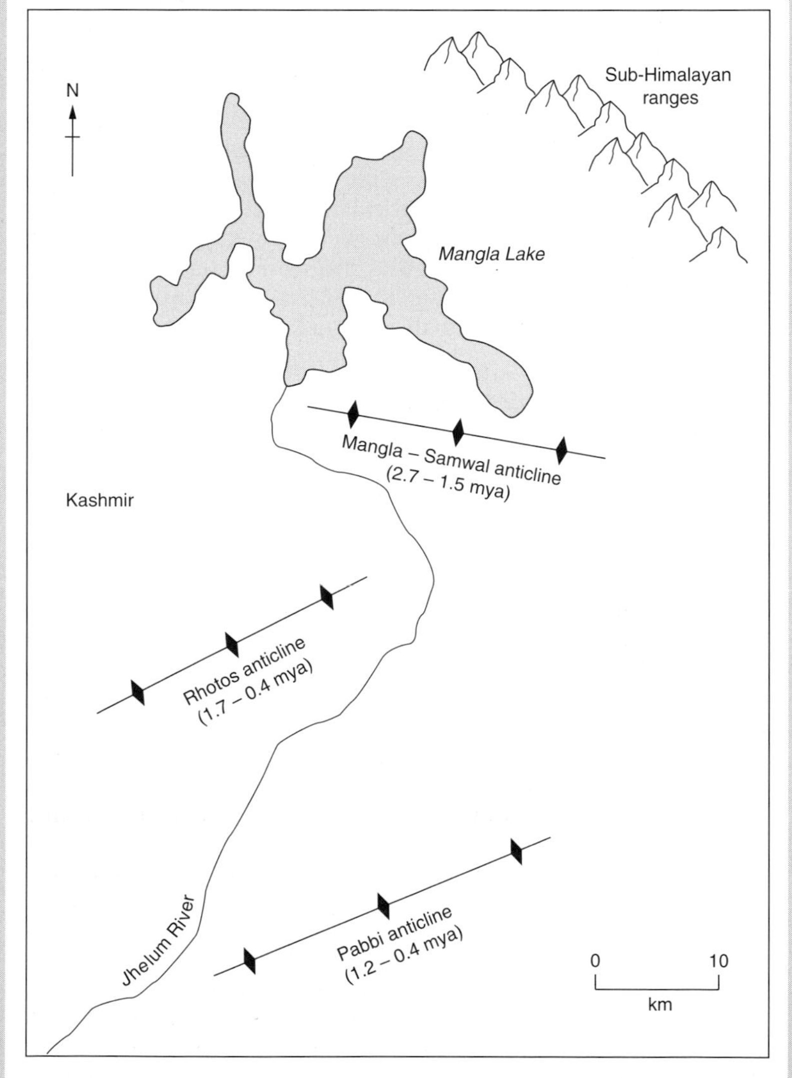

Figure 3.5 Young anticlines in the north-western Himalaya.

Phuntsoling is a border town in the southern edge of Bhutan, where it neighbours Sikkim in India. The Torsa river, an indirect tributary of the Brahmaputra, flows south from the Himalaya towards India and Bangladesh, leaving Bhutan through Phuntsoling. Along the valley of the Torsa there are many debris cones measuring only hundreds of metres across, deposits created from sediment that has been washed off the neighbouring slopes. These fans are unconsolidated material and are therefore very vulnerable to movement. Many of these fans have small terraces cut into them, and they are connected by numerous scarps, only a few metres high. Each scarp represents activity along a different small fault line within the solid geology of the Torsa area.

The Himalaya are one of the most beautiful mountain landscapes in the world. They owe their existence to a range of geological structures and processes that have combined to produce a collection of scenery and landforms without parallel. The many features within the mountain range are due to the variety of scales on which the forces creating the scenery have acted, from the mega-scale activities of the Indo-Eurasian collision zone to the micro-scale faulting activity that creates terraces on debris cones in Bhutan.

Summary Diagram

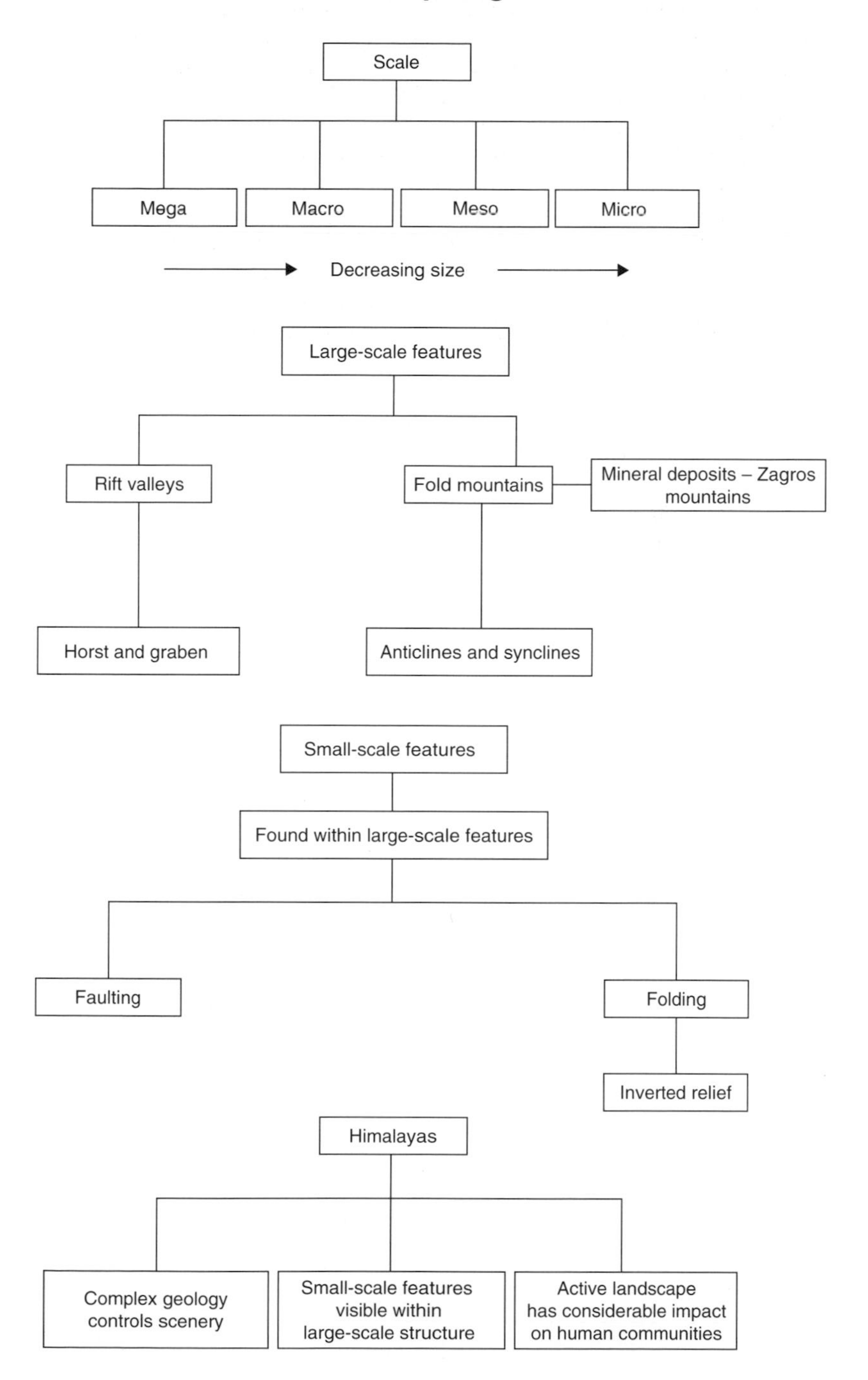

Questions

1. Why is scale such an important consideration in geography?
2. Define the following features of the Earth's surface:
   a) fold mountains
   b) rift valleys
   c) faults
   d) folds.
   Outline the forces that create each of these features.
3. Explain the development of a fold.
4. Discuss the impact that the building of fold mountains, such as the Himalaya, can have on the following aspects of geography:
   a) climate and weather
   b) drainage
   c) vegetation
   d) settlement
   e) economic activities.
5. Examine the impact that faulting can have on the landscape.

4 Weathering Processes and Features

KEY WORDS

Weathering: a change in the chemical, mineralogical and physical properties of rock in response to environmental conditions in a given area.
Erosion: the breakdown and removal of rock by the same force.
Mechanical weathering: a physical process that breaks down rock without causing any chemical change.
Chemical weathering: a process that breaks down rock by altering its structure through a chemical reaction.
Denudation: the lowering of the ground surface by a combination of all weathering and erosion processes.
Peltier curves: a set of graphs showing the relationship between climate and the amount of weathering likely to occur.

1 Weathering: An Introduction

Any rock that is close to the surface of the Earth is vulnerable to the forces of **weathering**. These forces are created as a consequence of complex interactions between the **lithosphere**, the **hydrosphere**, the **atmosphere** and the **biosphere**, and can affect rocks in a range of ways.

Chemical weathering comprises a group of processes that alter the chemical structure of the rock such that the physical properties are also changed. **Mechanical weathering** consists of the application of a physical force to rock such that the composition of that rock is broken up. A third category of weathering is sometimes given: **biological**. The action of plants and animals, however, can invariably be incorporated into either chemical or mechanical weathering either through the action of organic acids and other by-products of life, known as biochemical weathering, or through the physical impact of plants and animals through root system or burrows, known as biomechanical weathering, for instance.

Chemical and mechanical weathering processes are inextricably linked. As mechanical methods break rock up they expose an ever-increasing surface area of that rock to attack by water, the most significant of the chemical agents. Imagine a cube of rock with a face area of $4\,m^2$, providing a total area of $24\,m^2$. If mechanical weathering processes break this into eight equal cubes with faces of area $1\,m^2$ the total rock surface exposed will be 48 m^2, a doubling of the area vulnerable to attack by chemical weathering. Equally, as chemical weathering begins the arduous process of breaking down a volume of rock, structural weaknesses will be created, providing the very flaws that will be exploited by mechanical processes.

Finally, it is important to distinguish weathering from erosion. Weathering is the breakdown *in situ* of a volume of rock.

Erosion is carried out by the mobile forces of wind, water and ice, and causes breakdown of the rock and its removal by the same process. A weathered rock may be transported immediately weathering has taken place, as would happen, for instance, high on a cliff, but the transportation will be carried out by a different force to that which caused the weathering.

Denudation is a general term for the lowering of the landscape by any process, and therefore usefully includes all processes of weathering, erosion and transportation.

2 Mechanical Weathering

The processes included under the heading mechanical weathering are numerous, and, as Summerfield (1991) notes, "[Mechanical weathering] encompasses a range of mechanisms the relative effectiveness of which are not accurately known but clearly vary significantly as a function of environmental conditions". Therefore, in order to understand the changes the landscape experiences, it is necessary to comprehend the nature of the processes and the conditions in which they are likely to be most effective, either individually or in combination. This section will address the processes; other chapters address the issue of environments that favour particular forms of weathering in more detail.

a) Freeze–thaw weathering

Sometimes known as **frost shattering**, freeze–thaw weathering is often experienced first-hand by people living in temperate environments through the domestic inconvenience of burst water pipes. Pipes burst by the water inside the pipe freezing during subzero temperatures, expanding and forcing the pipe to expand until it fractures. The fundamental force at work is that of water freezing and expanding by 9.05% as it does so. This process occurs in any location where there is a fluctuation in temperature around 0°C, so the world's high latitude areas and high altitude areas are susceptible to significant weathering by this process.

The process is simple; water enters a joint or fault in the rock when the temperature is above freezing point, expands as it freezes when the temperature drops sufficiently and exerts pressure on the rock.

It is improbable that one cycle of freeze and thaw will be enough to fracture the rock, so the process only really occurs when there is an oscillation around freezing point, as shown in Figure 4.1.

Thus, the most potent circumstances for freeze–thaw weathering are those where there are frequent changes in temperature, since the frequency of the cycle is more significant than either the duration of the freeze or the depth of it. As a consequence, marine Arctic and Alpine environments would seem to be at greatest risk from this type of weathering. Recent academic research suggests that moisture content of the rock is important. If the rock is not saturated, the process of freezing will force water into the pore spaces in the rock, thus reducing the build up of pressure required to weather the rock.

The product of freeze–thaw weathering is usually large angular blocks of rock. The angle of the slope of the parent rock will contribute to the landscape formed. **Scree slopes** will develop on steeper slopes, such as are found in many British upland areas, such as Wast Water in the Lake District, while flatter terrain will generate boulder strewn areas known as **blockfields**, or **felsenmeer**. These are common at high latitudes, where the process can easily take place on a daily, or **diurnal**, cycle for much of the year. Many **glacial arêtes** have been sharpened and steepened by freeze–thaw action, to such an extent that they make for hair-raising walking terrain, such as that on Striding Edge in the English Lake District or in the Black Cuillins on Skye (see Figure 1.3). The boulders generated by freeze–thaw action on a glacial arête will often be found several hundred metres

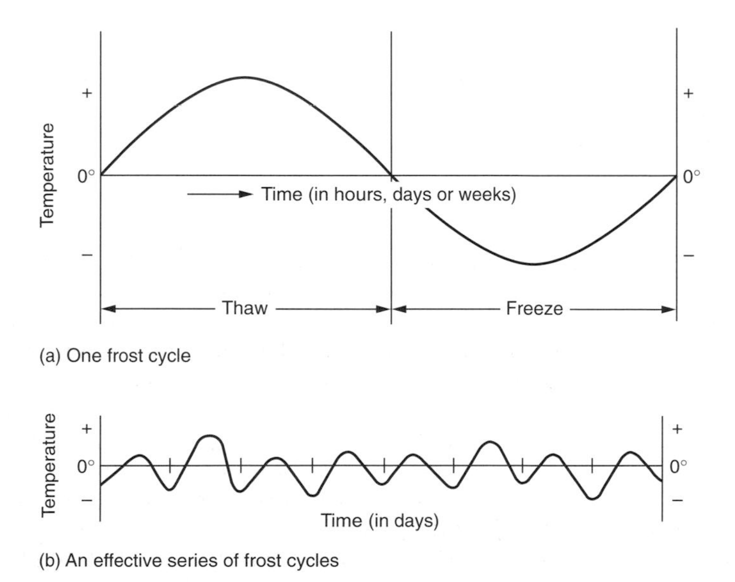

Figure 4.1 Frost cycles. *Source*: Collard (1988).

lower down, on scree slopes at the base of the corries bordering the arête.

Block separation is a common product of freeze–thaw weathering. It occurs most commonly in a rock with a clearly defined structure of joints and bedding planes. Limestone fits this description and mechanical weathering processes can attack these weaknesses easily. This can give rise to angular debris and landscapes, such as **mesa and butte** scenery, or **tors**. The regular joints in basalt can also lead to block separation, and the product of that can be seen in the Giant's Causeway in County Antrim, Northern Ireland, where the removal of some blocks has left the distinctive scenery that can be seen today. Freeze–thaw weathering is a process that can exploit such regular jointing and one that, therefore, often produces block separation.

b) Salt weathering

Although there are many clear parallels between freeze–thaw weathering and salt weathering, or salt crystallisation, there are also significant differences. Salt weathering is another process where rock is broken up as a consequence of pressure exerted on it from within the rock structure, but the environments likely to be most at risk from it could not be more different. Also, salt weathering operates at a much smaller scale than freeze–thaw in terms of the size of particle generated as an end product. There are also hints of an overlap between mechanical and chemical weathering in salt weathering because the process is caused by a chemical, even though it exerts a clearly physical, force.

All water contains minute quantities of **salts**, most commonly **sodium chloride**, although **sodium sulphate** is thought to be the most effective salt because of its particular solubility properties. Once water within the rock structure becomes saturated with salt, either through evaporation or temperature change, some of that salt will be forced to crystallise and will begin to exert pressure on the rock. This process, as with freeze–thaw weathering, will repeat itself, and the salt crystals will grow. Over time grains of sand can be dislodged or soil can be forced upwards and outwards, sometimes creating **patterned ground**.

As might be expected, salt weathering is commonest in **coastal locations** and in **arid areas**. The coastal deserts such as the Atacama (South America) and the Namib (South Africa) are two areas that experience a great deal of this type of weathering. Areas close to enclosed seas in arid areas, such as the Red Sea, are also vulnerable. These areas share the two key requirements for salt weathering: a ready availability of salt water, and high temperatures for evaporation to cause that water to become saturated with the salt, leading to crystal growth. Salt weathering also can occur in temperate climates, although the scale of the resulting features will be smaller.

c) Insolation weathering

Insolation weathering is the breakdown of rock as a result of successive changes in volume brought about by **heating** and **cooling** cycles. Theoretically, this works in a manner similar to frost shattering, except that it has proved harder to ascertain for sure that the forces applied are sufficient to shatter rock. In some desert environments, the diurnal temperature range can be up to 30°C, and this change in temperature can lead to the expansion and contraction of the surface layers of the rock. This creates stresses between those surface layers and the interior of the rock that may lead to fracturing or shattering. Laboratory experiments in the 1930s subjected small samples of different rocks to repeated heating and cooling without causing any disintegration, but it seems likely that if rocks are unable to expand in some directions, as is the case in reality, some breakdown will occur if sufficient forces are applied.

Laboratory work has also suggested that the addition of **water** to the process of insolation contributes to the quantity of disintegration experienced. Studies of buildings in arid areas have added to this theory; ancient monuments in Egypt, such as the Pyramids, show little signs of decay unless they are located close to the Nile river. **Fog** may also provide enough water to enhance the process of insolation. In the Namib coastal desert there may be fog on up to 100 occasions in a year, and this water may be sufficient to weaken some minerals in the rock enough to allow disintegration through a combination of chemical and mechanical processes. The fact that desert environments are often littered with shattered rock lends credence to this theory.

Insolation weathering can function at a smaller scale as well. The product of this process can be the granular disintegration of the rock concerned. This is caused by the differential weathering of minerals within the rock structure and causes the rock to break down into grains of sand. If granite is exposed to insolation weathering the dark crystals of mica will heat up more that the paler feldspars and quartz. This will set up stresses within the rock structure at a much smaller scale, and can lead to the breakdown of the rock crystal by crystal.

d) Pressure release

Although not strictly a form of mechanical weathering, pressure release is inextricably linked to this topic. When a great mass is removed from a layer of rock there is inevitably a change in the stresses within that rock. The removal of the pressure often leads to the expansion of the previously stressed rock, causing joints to open. This process is known as pressure release, or **dilatation**, while the joints are known as **pseudo-bedding planes**. Dartmoor shows evidence of such bedding planes. This is not surprising when the fact that the granite batholith that now forms the surface is thought to have been formed some 6 km below the surface. The subsequent removal of 6 km of other rock has allowed bedding planes to expand and new ones to form, making the granite that now forms the surface more vulnerable to other forms of weathering and erosion. This increase in susceptibility to other processes is probably the most significant consequence of pressure-release weathering. The product of pressure release is **exfoliation**, the peeling off of thin rock sheets. It is most likely to happen in a rock with many clear horizontal bedding planes, where the main weaknesses, therefore, run parallel to the surface. Thus, as well as pressure release, it is also likely that insolation weathering could cause exfoliation. Some spectacular examples can be found, especially in granite or sandstone. The Yosemite half dome in the Sierra Nevada in the USA is a fine example, which receives more detailed attention in the case study on page 73.

3 Chemical Weathering

Chemical weathering is caused by a reaction between the **rock** and the **air**, **water** or other **solutions** with which it comes into contact. This reaction need not involve all of the rock's constituent minerals. The end product can be any one of a range of reactions, from a change in colour to the total decomposition of the rock. It differs from mechanical weathering because mechanical weathering will lead to the breaking up of the rock into smaller segments, while chemical weathering causes the rotting down of the rock. There is some **overlap** between chemical and mechanical weathering. It is hard to decide whether salt crystallisation is a chemical or mechanical process because the processes that cause the salt to crystallise are chemical, but the forces it applies to the rock are physical.

The main processes of chemical weathering are **solution**, **carbonation**, **hydrolysis**, **oxidation** and **hydration**. It is important to remember that these processes do not occur in isolation. Many of them rely on water to occur at all, and since that is the case it is perfectly possible that more than one of the processes will happen simultaneously. Alternatively, the action of one process may create a mineral that is more vulnerable than previously to the action of a different process.

a) Solution and carbonation

Solution is a simple process that will be familiar to all of us. It is the dissolving of a soluble mineral in water, and occurs every time you stir sugar into a cup of tea. The natural world process takes a good deal longer, but provides the opportunity for flowing water to dissolve and remove minerals from the surface.

Few minerals are soluble in pure water, but the creation of a weak acid will allow solution to occur readily. The addition of naturally occurring **carbon dioxide** to **rainwater** in the atmosphere or soil turns water into a weak **carbonic acid**. Calcium carbonate is particularly vulnerable to solution in carbonic acid, meaning that limestones are prone to higher levels of solution than other rocks. When minerals in rocks are attacked by carbonic acid, the process of solution is known as carbonation. The process is slightly more complex than at first sight, because it occurs in two stages. Put chemically, this is expressed as follows:

$$CaCO_3 + H_2CO_3 \rightarrow Ca^{2+} + 2HCO_3^-$$

Calcite ($CaCO_3$) in the rock is transformed by the rainwater (carbonic acid, H_2CO_3), producing calcium ions (Ca^{2+}) and bicarbonate ions ($2HCO_2^-$), both of which are soluble and therefore dissolve in the water.

This action is not restricted to naturally occurring water. Any acid will attack calcite, and industrial pollution of the nineteenth and twentieth centuries may well have hastened the landscape evolution in some areas, notably the Carboniferous limestone of north Yorkshire. This area provides a spectacular example of how this process works and the unique landscapes that can be produced, and the nearby industrial heartlands of west Yorkshire and Lancashire may have contributed. This location is investigated in more detail in the case study on page 70.

b) Hydrolysis

Hydrolysis occurs when water reacts with minerals in the rock and **decomposes** them. It is a common process of chemical weathering, not least because the minerals that are vulnerable to it include **silicates** such as feldspar that are found in many granites. The hydrogen ions in the water react with metal cations in the mineral structure causing a complex series of changes that may ultimately convert the silicate mineral into a clay mineral.

The precise chemical nature of the silicate mineral will dictate the outcome. Perhaps the best-known example of hydrolysis in Britain is the creation of kaolin in the granites of the south-western batholith. Kaolin is the product of the hydrolysis of potassium feldspar, or **orthoclase**:

$$2KAlSi_3O_8 + 2H_2O \rightarrow Al_2Si_2O_5(OH)_4 + K_2O + 4SiO_2$$

This complex equation is more easily understood in words:

orthoclase + water → kaolinite + soluble potassium oxide + soluble silica

The soluble products of this reaction are dissolved away, leaving the kaolinite behind.

However, kaolin is not the most common clay mineral produced by hydrolysis. Illite and montmorillonite are more common, although less useful than kaolin. Kaolin is created from Dartmoor granite because the presence of potassium in the feldspar leads to the breakdown of the whole rock, since the removal of the feldspar leaves unconsolidated mica flakes and quartz crystals. As the feldspar turns to kaolin it also expands in volume and the expansion contributes to this physical disintegration.

c) Hydration

Hydration is the chemical equivalent of saturating a sponge. Some minerals can absorb water into their structure and in doing so experience significant changes. This can have minor effects such as colour change, but expansion of the rock as water is absorbed is also a consequence. This is made more important because the hydration reaction is reversible. Thus, these volume changes can be repeated and make the rock vulnerable to mechanical failure such as cracking or granular disintegration. In addition to these changes, hydration also assists other forms of chemical weathering by introducing water throughout the crystal structures. The mineral anhydrite (calcium sulphate) is vulnerable to hydration, and it becomes gypsum, which is used in the manufacture of plaster of Paris:

$$CaSO_4 + H_2O \rightarrow CaSO_4 \cdot H_2O$$

d) Oxidation

Oxidation is a process with which we are all familiar. Rusting of metals is an everyday sight, and it occurs just as readily in the natural world. Iron, manganese and sulphur all occur with abundance in minerals, and the reaction between them and oxygen dissolved in water is oxidation. This process commonly causes colour changes in minerals and often explains any reddish-brown tinge to exposed rock faces or soils. In addition to this relatively cosmetic change, the chemical changes wrought by oxidation can render a mineral more vulnerable to another form of chemical weathering. Minerals that have been oxidised also increase in volume. This applies stresses to the rock's physical structure and can lead to more rapid mechanical weathering than would otherwise have been the case.

4 Biological Weathering

Biological weathering is probably the most difficult of the three weathering types to define. Each type of weathering that could be included in the category of biological weathering could also just as easily be classified as either mechanical of chemical weathering. This is because biological weathering is best defined as any sort of weathering that is carried out by a **living organism**, or by a by-product of a living organism. It is best divided into **biomechanical weathering** and **biochemical weathering**.

a) Biomechanical weathering

Biomechanical weathering is mechanical weathering that is carried out by a living organism. The action of tree roots on jointed rocks can be classified as biomechanical weathering. This is a common method of mechanical weathering in areas where bedrock is close to the surface, and the climate permits the growth of substantial trees. It is also commonly seen in urban areas as tree roots force their way through tarmac and concrete in search of water. Whatever surface or rock is affected, the action of tree roots opens passageways through the rock or tarmac that water can pursue. This increases the vulnerability of the land to other forms of weathering, as well as the distortion that has been caused by the direct action of the tree roots.

As well as tree roots, more dynamic organisms can disturb the landscape and open pathways for water. Burrowing animals of all kinds create easy routes through soil layers to the bedrock for water. Evidence of this can often be found in areas that are rich in such animals, such as chalk uplands. Burrows and scrapes can expose the bedrock directly if the soil is thin enough, and provide a channel through the soil if it is sufficiently thick.

b) Biochemical weathering

Biochemical weathering is perhaps the more important of the two categories of biological weathering. A great deal of bedrock, especially in temperate and tropical climates, is covered by a layer of soil and other organic matter. As organic matter decays, it creates a variety of chemicals, including a range of organic acids. As water passes through the organic layer of the soil, some of these organic acids are dissolved into the water. This can make the water, already mildly acidic after its passage through the atmosphere, a more effective agent of chemical weathering once it reaches the rock. This process is part of **chelation**, a complex biochemical process that includes the breakdown of soil as well as the eventual weathering of rock. Although the processes that break down the rock are ultimately those described in Section 3 of this chapter, the fact that their efficacy has been enhanced by the presence of organic acids makes the inclusion of chelation in the category of biochemical weathering worthwhile.

5 Factors Affecting Weathering

All the weathering processes outlined in this chapter are affected by a variety of other influences. The rate at which they happen and the outcome they produce can vary according to the conditions in which they occur. Rock type is one of the factors that influences the rate and product of weathering, but that is discussed in more detail in Chapter 2. This section will concentrate on the impact of **climate**, **soil** and **topography** on weathering processes and their outcomes. These are just three of the factors that determine the rate and type of weathering, as shown in Figure 4.2.

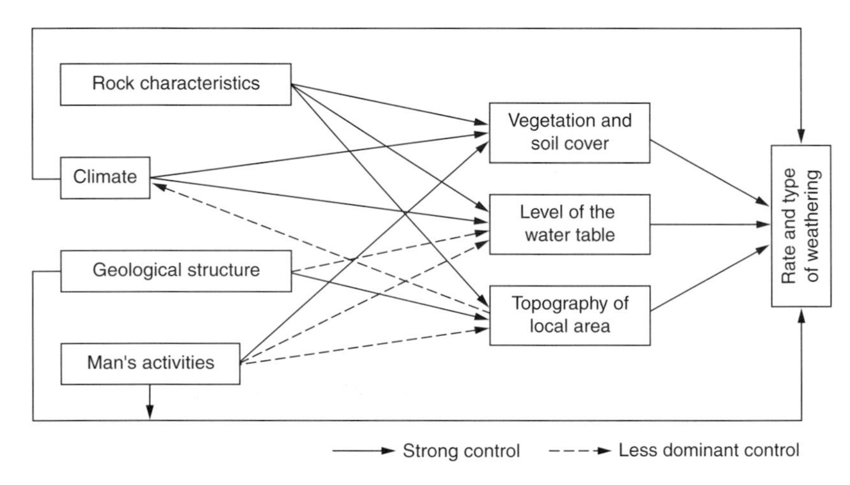

Figure 4.2 Factors affecting aspects of weathering. *Source*: Collard (1988).

a) Climate

Climate is without doubt one of the most important controlling factors of the nature of weathering that will occur in any given location. This is not especially surprising, given that virtually every weathering process requires either the addition of **water** or a change in **temperature**, or both, to work effectively. Changes in temperature have a significant impact on the rate at which **chemical weathering** processes will occur, and the addition of water can make a difference to the rate at which **mechanical processes** function. The relationship between temperature, rainfall and weathering is a complex one, and a great deal of work has been done by geomorphologists to try and unravel these complexities. In 1950 **Peltier** devised a series of curves to illustrate this relationship. The curve for **frost action** (Figure 4.3) shows that the optimum environment for frost shattering is not the coldest environment, but the one that is likely to produce the greatest number of frost cycles. Equally, the efficacy of frost weathering

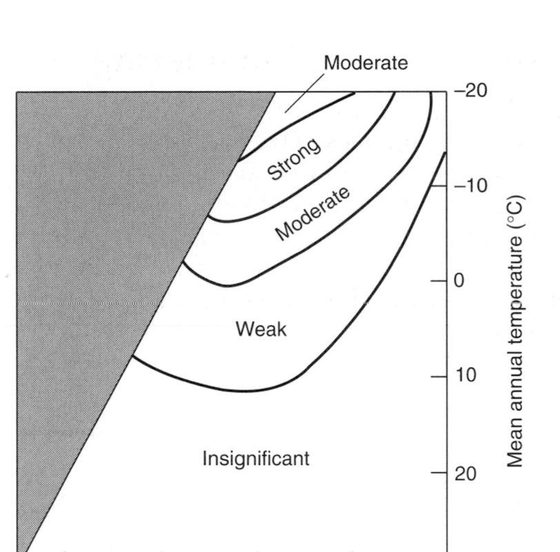

Figure 4.3 Peltier curve for frost action. *Source*: Collard (1988).

decreases with rising temperature and also with decreasing precipitation.

Similarly, Peltier's curve for **chemical weathering** (Figure 4.4), shows that the environments that will favour the most effective chemical

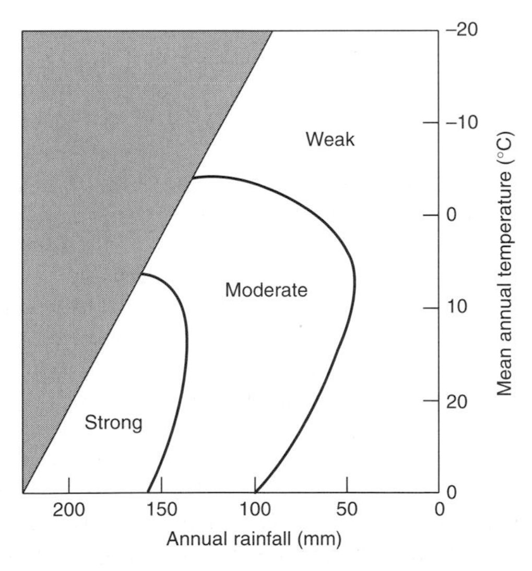

Figure 4.4 Peltier curve for chemical weathering. *Source*: Collard (1988).

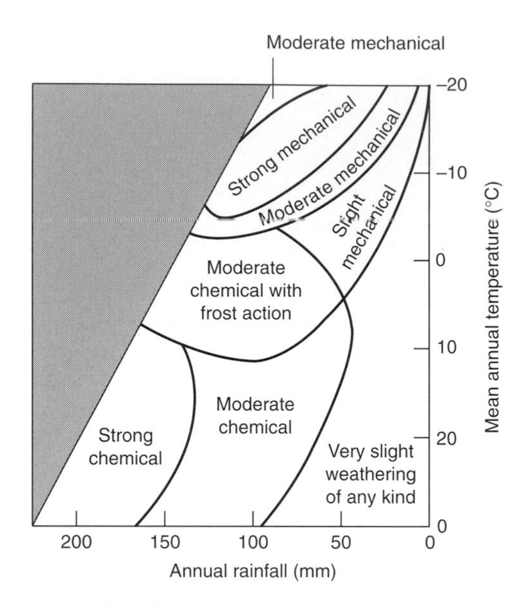

Figure 4.5 Peltier curve for weathering regions. *Source*: Collard (1988).

weathering are those that provide the hottest and wettest conditions in which the processes can function. Hence, tropical environments should experience the greatest rates of chemical weathering while undergoing very little mechanical weathering.

Finally, Peltier produced a general graph (Figure 4.5), designed to combine his other work and illustrate the overall impact of climate on weathering. Here it can be seen that temperate areas such as the British Isles might be expected to experience moderate chemical weathering with frost action, while colder areas will not notice any chemical weathering of any kind. As conditions become hotter and wetter, the amount of chemical weathering increases at the expense of mechanical weathering, while hot and dry environments experience only the smallest amounts of weathering.

Peltier's ideas are not perfect; he seems to have overlooked the impact of **thermal expansion** as a weathering force, but his work on chemical weathering certainly provides a good reference point for the differences between tropical and temperate weathering. An example of this is shown in detail in the case study comparing Malham and Ankarana.

Tropical weathering

The rate of chemical reactions increases with **temperature**. This means that the rate at which bedrock will be weathered in tropical environments will be much quicker than in temperate environments.

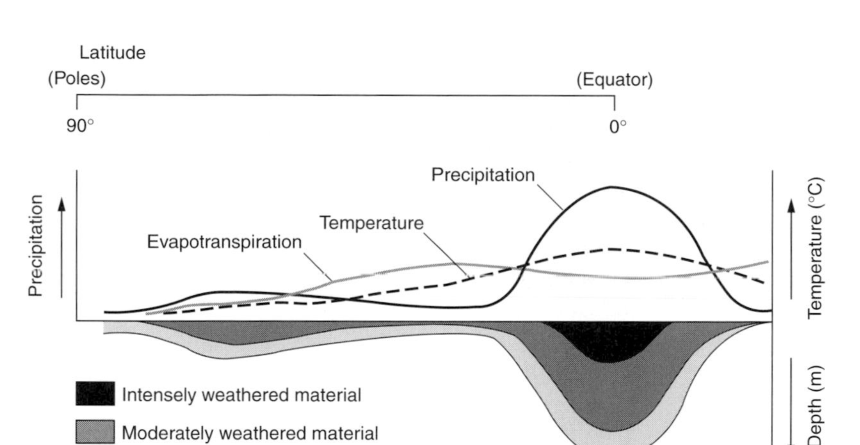

Figure 4.6 Latitudinal variations in weathering depth.

Consequently, the depth to which weathering will extend is greater in tropical environments, because the processes are able to break down the rock quicker and therefore penetrate deeper. Figure 4.6 shows this clearly; as the amount of precipitation in particular increases, so does the depth of the weathered material.

The reasons for the greater amount of weathering in tropical areas are mainly chemical. Silica, a very common mineral, is more soluble at higher temperatures, and the quantity of rainfall will leach away other minerals more than in drier environments. The amount of rainfall means that the bedrock is saturated more often in tropical environments, meaning that hydration is a more effective process there than in drier locations.

Tropical processes are relevant when explaining the differences between karst landscapes in different parts of the world, as discussed in Chapter 2.

Arid weathering

While tropical areas undergo intense weathering as a result of a hot and humid climate, hot arid zones experience a hot and dry climate. However, the diurnal cycle of temperatures in arid zones can produce significant quantities of weathered material not alluded to in Peltier's work. **Thermal fraction** can breakdown the bedrock very effectively as a result of insolation weathering. The lack of vegetation means that much bedrock is exposed in such areas. Rock surface temperatures can reach 75°C during the summer months, and experience a diurnal air temperature range of 25–30°C. This creates patterns of expansion and contraction that can disintegrate the rock. In mountainous areas, frost shattering can occur. Dew can provide the source of water and temperatures in the Atlas Mountains of north Africa, and the Andean parts of the Atacama desert, frequently reach freezing point. Dew is not the only source of water in such areas; Oukaimeden in the Moroccan Atlas has a regular, albeit short, skiing season.

No soil to speak of forms in arid areas, because of the lack of organic matter, and rates of weathering tend to be so slow that weathered material has no time to accumulate before being removed. Hard crusts known as **duricrusts** are a common feature of desert rock surfaces, and provide an additional layer of resistance to weathering. These are accumulations of salts deposited as a consequence of the high evaporation rates in arid areas.

Periglacial weathering

Periglacial areas are those close to the fringes of ice sheets, where the ground remains frozen for all or part of the year. Repeated freezing and thawing creates a unique set of landforms, and weathering processes are prominent in the production of these landscapes. Unsurprisingly, freeze–thaw action is the most common process in such areas, and creates the angular, unsmoothed debris that is commonly found in periglacial environments. Scree slopes such as those shown in Figure 4.7 are an example of this. Individual rocks are also

Figure 4.7 Scree slope on Jebel Toubkal, Morocco.

vulnerable to freeze–thaw weathering, and the granular disintegration of boulders as water penetrates between the grains or crystals in the boulder is a well-documented feature. Freeze–thaw action seems to be most effective in areas with no snow cover. The insulating properties of a layer of snow mean that temperatures below such layers are kept higher, thus reducing the number of freeze–thaw cycles, and the amount of weathering that takes place. It might be reasonable to expect rates of chemical weathering to be low in periglacial areas because water is stored as ice for much of the year, the temperatures are low and there is little biotic activity. This has not been proven to be the case, and it is thought that much evidence of chemical weathering might have been removed by highly effective glacial processes during the Pleistocene glaciations.

b) Soil

The depth and nature of **soil cover** in a given area will have a significant impact on the amount of weathering that can affect the rock underneath. If a mechanical weathering process such as frost shattering or insolation weathering has led to the creation of a layer of soil, then that soil will eventually prevent the process from taking place by protecting the rock from the extremes of temperature that are needed for the process of weathering to function. Soil is an effective insulator, but it can also promote other forms of weathering. Organic acids in the soil can cause **chelation** and carbonation to be enhanced, thus increasing the amount of chemical weathering that the rock below may experience. In turn this weathering may lead to further development of the soil, allowing more plants to grow and further increasing the amount of organic acids available to the water. In tropical areas, deep soils and regolith, as shown in Figure 4.6, may help explain why Peltier's diagram shows little or no mechanical weathering in areas that might be thought to be susceptible to insolation weathering. The thin layer of soil protects the bedrock from the direct action of the sun, thus reducing the opportunities for mechanical weathering in such an area.

c) Topography

The result of weathering processes has to be considered with reference to the environment in which these processes occur. Much of this is bound up with slope and mass movement processes that will be further considered in Chapters 5 and 6, but some basic points can be made here. For weathering processes to continue unchecked they need to work on fresh rock once material has been broken down. Thus, some transport processes are needed for the weathering processes to be able to access new material. If the **slope** on which the weathering occurs is too gentle, then weathered material may accumulate *in situ* and prevent the access. Too steep a slope, and there is likely to be a shortage of water at the weathering site because of the rapidity with which water would flow down such a slope.

The **altitude** of a location will also have an impact on the effectiveness of processes that will occur in that location. In mountainous areas, an altitude above the tree line is likely to be a location that has an increased chance of effective freeze–thaw action because of the reduced insulation from vegetation and soil. Too high an altitude will decrease the number of diurnal cycles that cross freezing point.

Aspect also has an impact, especially on temperature-dependent processes such as insolation weathering and freeze–thaw action. A deep, glaciated valley at a high latitude, with a north-facing slope and a south-facing slope, will experience a range of conditions, some of which will promote weathering more than others. Areas that receive no direct sunlight at all, such as north-facing slopes (in the northern hemisphere), may not experience much freeze–thaw action if the temperatures are below freezing point. Equally, areas that receive regular insolation, such as the upper slopes of a south-facing slope, may not experience many freeze–thaw cycles because the temperature is maintained above freezing point. The lower part of the south-facing slopes may receive insolation during the summer months only, meaning that for part of the year the location has no freeze–thaw action. All of these scenarios are dependent on the air temperature. If the insolation is needed for the freeze–thaw cycle to occur, then the southern slopes will experience more weathering, but if the insolation prevents the freeze–thaw cycle occurring then the northern slopes may show more evidence of weathering.

CASE STUDY: LIMESTONE WEATHERING IN MALHAM AND ANKARANA

Karst landscapes have been considered in Chapter 2. This case study takes that consideration further by comparing two areas with very similar bedrock, Limestone, in very different climate belts. Malham, in north Yorkshire, is around 400 m above sea level and receives some 1100 mm of rainfall each year. The Carboniferous limestone in this area is between 330 and 345 million years old, and was subject to glaciation in the Pleistocene ice ages. Its location contrasts with that of the Ankarana Special Reserve in northern Madagascar.

Ankarana also has limestone as its bedrock, although from a later geological time, the Mesozoic. Despite its greater age, it has very similar characteristics to the Carboniferous limestone in north Yorkshire. However, the climatic parameters are very different. Ankarana has annual temperatures ranging from 25 to

30°C and a seasonal tropical climate. It experiences a distinct wet and dry season, with almost no precipitation during the dry season, but several hundred millimetres of rain can fall during a few weeks in the wet season. Annual rainfall is in excess of 2000 mm, throughout the area, but the high ground of the limestone within the Special Reserve can receive more.

Both areas have a protected status. Malham is within the Yorkshire Dales National Park, while the Madagascan government has decreed that Ankarana should have Special Reserve status, in many ways equivalent to a National Park in Britain.

Higher temperatures in tropical areas have two significant impacts on limestone weathering. The increase in the rate of reaction is the first, since it allows a greater degree of carbonation to take place. High temperatures also cause faster growth rates in plants, increasing the production of organic acids that have a major role to play in the weathering of limestone. Greater rainfall levels have a direct link with the amount of weathering, since the rain is the weathering agent. Annual rainfall figures for Malham (1100 mm) seem insignificant when compared with Ankarana's 2200 mm.

The amount of carbon dioxide in the water helps determine the ability of that water to dissolve calcium carbonate. This ability is known as the **aggressivity** of the water, and any water containing carbon dioxide, or another acidic agent, has this ability. Aggressive water will continue to dissolve calcium carbonate until it becomes chemically saturated, at which point the precipitation of calcium carbonate could occur. If more water crosses a given stretch of limestone then that limestone will come into contact with more carbon dioxide, increasing the amount of carbonation that can take place.

Limestone structure is particularly vulnerable to chemical weathering by water because it has a pronounced jointed structure, with joints and bedding planes as shown in Figure 4.8. Although the rock structure itself is impermeable, the presence of these joints and bedding planes means that a mass of limestone is permeable, since water can flow down joints and along bedding planes. Little vegetation grows in upland areas based on Carboniferous limestone, because the soil formed on such areas tends to be thin and not very fertile. Also, the terrain is not very hospitable to plant growth.

Limestone areas produce a range of landforms at all scales. Steep escarpments and spectacular valleys are common, although these owe more to the general structure of the rock than to climatic conditions. Smaller scale surface landforms often show characteristics that have formed because of the inter-

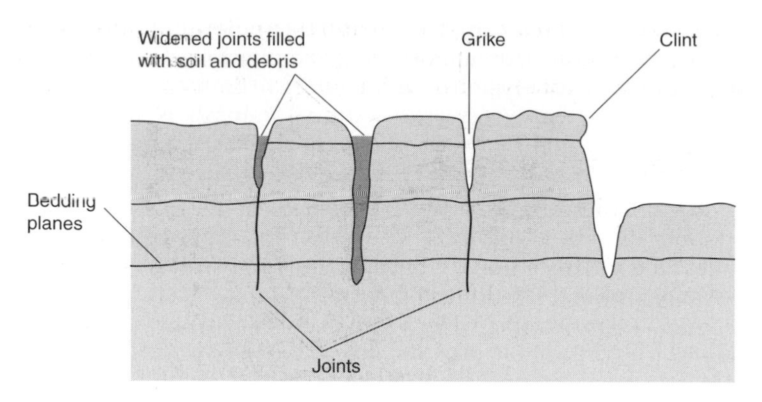

Figure 4.8 Structure of Carboniferous limestone.

action between the rock and the climatic type dominant in that area.

Small-scale temperate landforms

As a result of the characteristic bedded structure of limestone, plateaux are a dominant feature of many limestone regions around the world. Glacial scouring of the limestone has allowed carbonation to take place on newly exposed rock, forming the plateau of **limestone pavement** above Malham Cove in north Yorkshire. This generates the characteristic small **clint** and **grike** features shown in Figure 4.8, where water has percolated through the joints and widened them to form a vertical-sided crack between the blocks of limestone. Typically, the blocks in the Malham area are 1 m wide and up to 3 m long, while the grikes are up to 2 m deep and 15–20 cm wide. These clint and grike features bear some resemblance to more extreme features found in the Ankarana region. They are characteristic of limestone pavements found above Malham Cove, throughout the Ingleborough region of north Yorkshire and in south-west Ireland.

Not all the blocks found in a limestone pavement are as clearly defined as those in Figure 4.8. Some clints contain grooves and pans as well as clear-cut grikes between them. These features formed when the limestone was covered by soil that inhibited the movement of water and made the water more acidic. As the water could not move freely it created **runnels**, shallow grooves in the surface of the clint, which now show the route the water took as it flowed slowly into the joint between the clints. **Solution hollows** have also developed on the surface of some clints. These pan-like depressions were formed when water ponded under the soil, but above the limestone. The soil cover that allowed this variety of small-scale surface features to develop was present during Neolithic times, but has gradually been removed subsequently as a consequence of farming in the area.

Small-scale tropical landforms

Tropical limestone environments, such as those in Ankarana, produce a rather different landscape from that of temperate limestone environments, although the large-scale characteristics are often similar. Bedding planes and joints dictate the precise location of the land surface and of gorges and valleys through the Ankarana massif. It is only at the small scale that the landscape is so different. Levels of vegetation are also greater in Ankarana, especially in the gorges that break through the area, suggesting that rates of weathering may be further increased by the addition of organic acids.

The scale of the eroded features is much greater than in Yorkshire. The joints between the blocks of limestone have been eroded to a depth of 20–30 m and the top surfaces of the blocks have been sharpened to form **solution spikes** and **solution flutes**. The solution flutes bear a strong resemblance to the runnels found in Yorkshire in that they are formed as water follows the contours of the rock, except in conditions which lead to a much more dramatic landscape. Essentially this is a much more extreme version of the landscape seen in England; rather than breaking bones, this landscape is capable of skewering a person. The flutes provide a channel for the high levels of rainfall received during the rainy season and each successive deluge removes a little more calcite, enhancing the drama of the landscape. No human settlement could survive in this area, not just because of the terrain, but also because of the shortage of water during much of the year. During the dry season there is no precipitation, while the wet season sees floods and conditions just as difficult as the dry season droughts.

CASE STUDY: YOSEMITE VALLEY

The Yosemite Valley in the Sierra Nevada of California, USA is a good example of many processes of mechanical weathering and their impact on granite. The results are among the most spectacular scenery to be found on granite landscapes anywhere (see Figure 4.9).

Figure 4.9 Yosemite Valley scenery.

Geological background

During the Cretaceous period of geological time, about 100 million years ago, many **granitic intrusions** formed within the crust underneath what is now the Sierra Nevada. These intrusions coalesced to form a chaotic mosaic of granitic rocks, all with slightly different characteristics. Some of the intruding magma broke through to the surface and formed volcanoes, perhaps similar to mountains such as Mount St Helens and Mount Hood in the Cascade range of Washington and Oregon today (see Figure 2.1). By the late Cretaceous period, some 70 million years ago, the volcanoes had become extinct and had been eroded away by water, wind and ice. The granitic rocks that had cooled underground had made their way to the surface as the volcanic rocks were removed. The removal of the rocks above the intrusions released **pressure** that had previously been applied to the granites. This promoted the development of joints and bedding planes within the rock structure, at a local scale. This has been of great importance in the development of the landscape of the Yosemite area today.

Tectonic activity that began in the Cenozoic period (from 65 million years ago) tilted the block of intruded material so that the Yosemite region had a long, gentle slope from east to west, towards the Sacramento Valley in central California and the Pacific seaboard, and a steep west to east slope running towards the Great Basin of Nevada. This uplift was caused by the subduction of the Pacific plate underneath the North American plate, and superimposed a series of regional-scale joints on the Sierra Nevada granites. These joints have determined the location of major features within the range, such as the Yosemite Valley.

Processes of erosion

Weathering processes do not happen in isolation, and in order to identify the role of weathering in Yosemite some consideration must be given to the **erosive processes** that the area has experienced. **Water**, both liquid and solid, has attacked the rock, exploiting weaknesses at the regional and local scale.

As the Sierra Nevada area was uplifted by the Cenozoic tectonic activity so running water was provided with increasing amounts of energy with which to erode the landscape. These conditions allowed **canyons** to be cut through the western slopes of the Sierra Nevada. These deep, steep-sided river valleys divided areas of high ground from each other. The upland areas did not share the dramatic characteristics of the valleys, since they were not being attacked by the rejuvenated streams. The Yosemite Valley was cut during this time, by the Merced river.

During the Quaternary period **ice** became a dominant force in the erosion of the Sierra Nevada. It is uncertain how many times ice advanced and retreated across the area, but it is clear that at least three major glacial episodes occurred. These covered the highest parts of the range with ice caps from which protruded tongues of ice, or valley glaciers, which followed the pre-existing river valleys. Thus, the Yosemite Valley, originally cut by the Merced river, was then modified by the actions of ice. This had the effect of straightening and deepening the already substantial valley, creating a valley that is, in some places, 350 m deep. The glacial episode thought to be responsible for this erosion is known to American glaciologists as the Sherwin glaciation. This occurred 1 million years ago, and may have lasted for 300 000 years. Subsequent glacial episodes were not as substantial and did not fill the valley with ice. As a result the spectacular upland scenery such as the Half Dome and El Capitan, and the many spires and pinnacles, have had at least 700 000 years to form, since they are far enough down the western slope of the whole range to avoid covering by the ice caps, and high enough above the valley floor to miss the erosive force of the more recent valley glaciers.

Exfoliation domes

Granite develops patterns of cracks as it cools underground. However, these cracks can be enlarged by pressure release, which can also create new bedding planes, sometimes known as sheet joints. When the Sierra Nevada batholith was exposed during the late Cretaceous period the release of the pressure applied by the rock above it led to the development of a series of clearly defined bedding planes in the rock. These rocks then crack along the weaknesses leading to the peeling off of successive layers of rock, known as **exfoliation**. This has happened in the Yosemite area with spectacular consequences. Around Yosemite the tops of the batholith have been exposed in a series of domes, such as the North Dome and the Half Dome. As the pressure was released from these rounded bulges of granite, the layers began to peel off, providing the distinctive scenery. The weaknesses also provide opportunities for other weathering processes to act, such as freeze–thaw weathering. Scree slopes are a common consequence of such action and can also be seen in the Yosemite Valley.

Summary Diagram

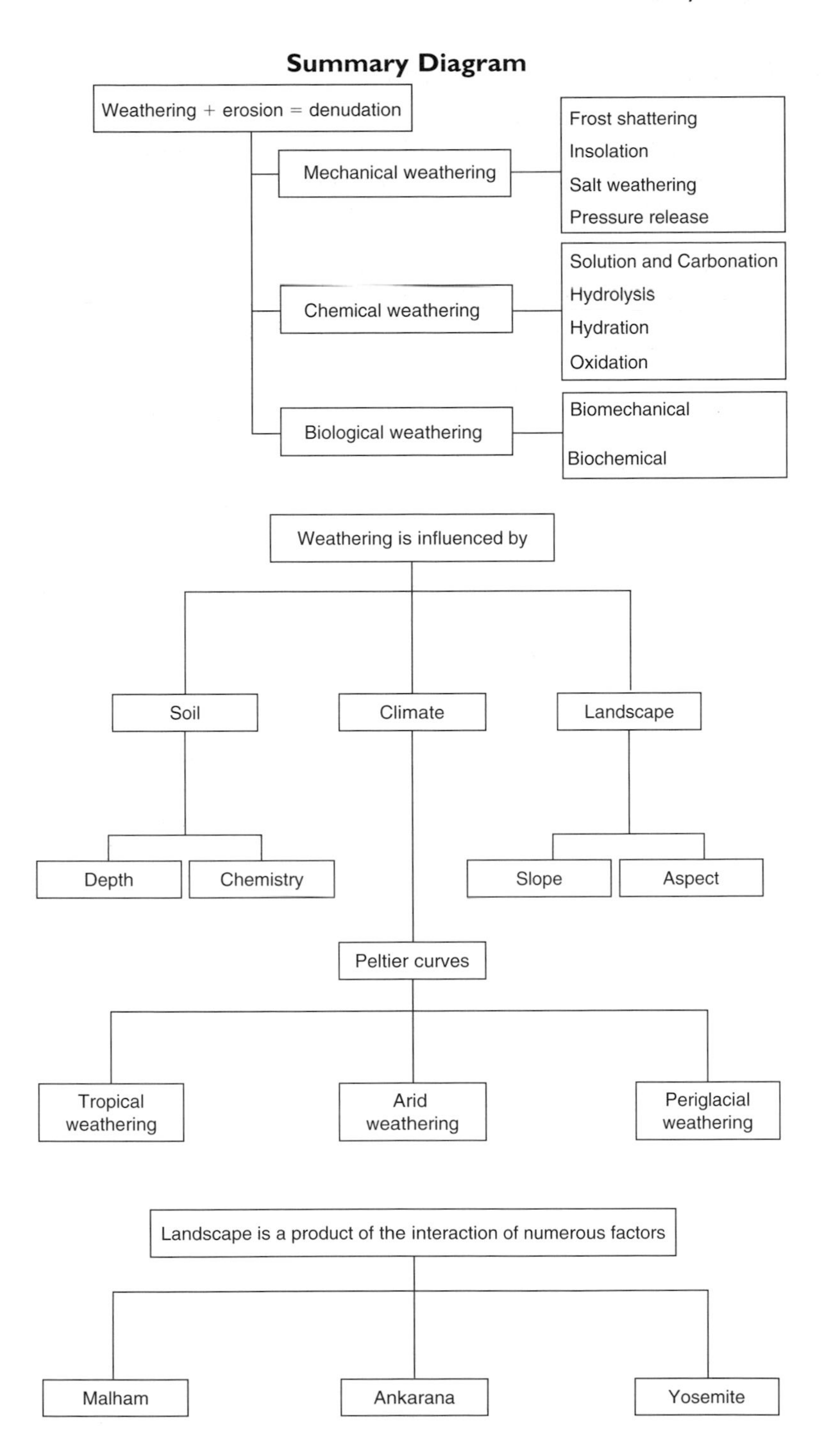

Questions

1. **a)** Explain how the aspect of a slope can influence the location of freeze–thaw weathering.
   b) Show how the answer to **a)** is dependent on latitude.
2. What factors affect the rate of **a)** chemical weathering and **b)** mechanical weathering?
3. Why is it not possible to clearly distinguish biological weathering from mechanical and chemical weathering?
4. Identify and explain the type of weathering that will dominate in each of the following environments:
   a) arctic
   b) desert
   c) tropical rainforest
   d) alpine.

5 Mass Movement

KEY WORDS

Failure: the sudden development of fractures in rock or soil and the inability to resist continued stress.
Flow: the downhill movement of earth or rock
Marginal land: land that is barely suitable for the human activity being carried out on it. It may be too steep, too wet or too dry.
Shear strength: the maximum resistance of a material to stress applied to it. When the stress applied to a material on a slope exceeds its shear strength a mass movement will occur.
Instability: a slope will become unstable and liable to mass movement when the shear strength of the material is exceeded.

1 Defining Mass Movement

Mass movement is the downhill movement of surface material primarily under the influence of gravity. Mass movement may be assisted by added buoyancy provided by water that has entered the material through precipitation or snowmelt. Distinguishing between mass movement processes and **transportational** processes is not always easy, since the divide between a mass movement process involving a significant quantity of water and a fluvial transportational process is not necessarily clear.

There are a number of ways of **classifying** mass movement processes, but the operation of all mass movement processes depends on the development of instability on a hill slope.

2 Classifying Mass Movements

Mass movements can be classified in a number of ways, two of which will be considered here.

a) Carson and Kirby's classification

Carson and Kirby (1972) proposed a classification based on the rate and moisture content of the mass movement. This can be illustrated on a ternary diagram, as shown in Figure 5.1.

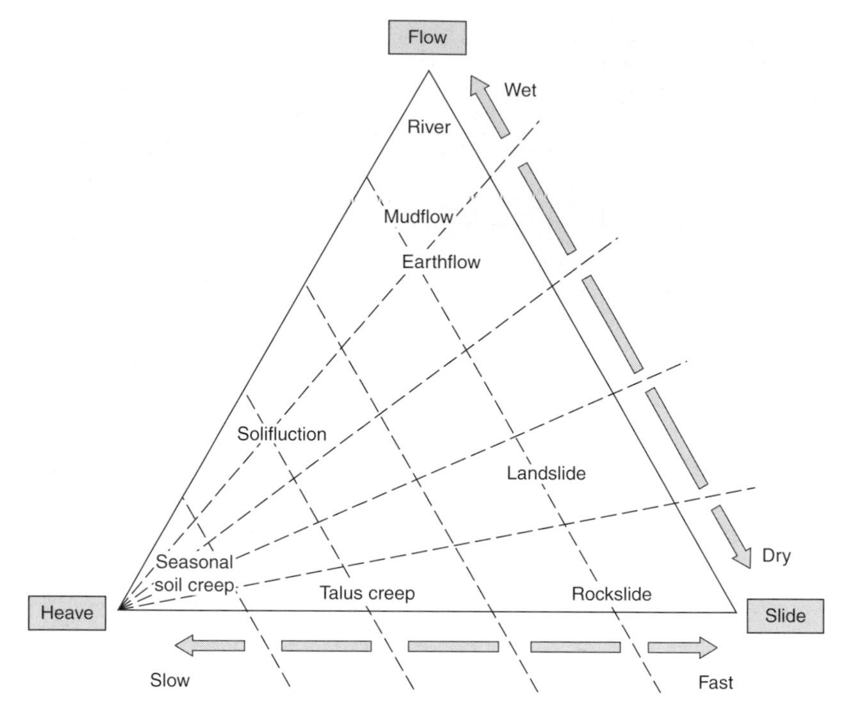

Figure 5.1 Carson and Kirby's classification of mass movements.
Source: Carson and Kirby (1972).

As Figure 5.1 shows, each type of mass movement can be plotted on the triangle according to its rate, measured along the base of the triangle, and its moisture content, shown along the right-hand side of the triangle. Thus, heaves are invariably slow processes, flows are wet processes and slides are dry.

b) Varnes classification

Varnes (1978) suggested a different style of classification. Rather than considering the nature of the mass movement as Carson and Kirby did, he based his arrangements on whether the movement was predominantly lateral or vertical. In other words, the movement would be classified according to whether the material spread outwards or dropped downwards. Within these two categories he then divided the movements according to their precise nature. A simplified version of Varnes classification is shown in Figure 5.2.

As with all **classifications** these two classifications of mass movements must be treated with some caution. Most mass movements are actually a combination of more than one of the processes listed in the classifications. This point applies to most classifications used in

Primary movement	Type of mass movement		Moisture content	Rate of movement	Materials involved
Lateral	Creep		Low	Extremely slow	Deformable rock or soil
	Flows	Solifluction	High	Slow	Soil
		Mudflow	Extremely high	Slow	> 80% clay sized particles
		Debris avalanche	High	Very rapid	Coarse debris
	Slide	Slide	Low	Variable	Rock mass or debris moving parallel to the slope
		Slump	Low	Variable	Soil or rock moving in rotation along a concave failure plane
	Heave		Low	Extremely slow	Widespread incremental downslope movement
Vertical	Fall		Low	Rapid	Rock or debris
	Subsidence		Low	Variable	All; cavity collapse (very rapid) or settlement (slow)

Figure 5.2 Varnes classification of mass movements.

geography, but they are often a very useful way of beginning a study of a subject such as mass movements.

3 Causes of Mass Movements

The operation of mass movement processes relies on the development of **instability** on a slope. If instability occurs then a mass movement process will result. Some mass movements are rapid, spontaneous responses to a slope failure while others are a more continuous process occurring over much longer periods of time. The rapid mass movements tend to occur when the stresses placed on the material in the slope greatly exceed their **shear strength** for a short period of time, while slower processes are more likely to occur when the stresses only just exceed the strength of the slope.

The major source of stress on a slope is **gravity**, and the magnitude of the gravitational force is a consequence of the angle of the slope and the weight of the sediment and material that the gravitational force is acting upon. The opposite forces, the resistance of the slope to gravity, can be derived from friction between the particles in the sediment, in the case of sands and gravels, or from cohesive forces between the particles. The latter situation occurs when the material is made from silt and clay particles. **Cohesion** is also influenced by moisture availability; too much moisture and the cohesive bonds are broken. Rock slopes generally have the greatest internal strength since the effects of solidification and crystallisation are to create very strong chemical and physical bonds.

Mass movement occurs when the stresses applied to the slope exceed the internal strength. Mass movements can be triggered because of an increase in the stresses, or a reduction in the strength. A period of **heavy rainfall** can trigger mass movement by reducing the cohesion between a previously stable slope made of clays and silts. The addition of a significant quantity of water in this way also increases the weight of the sediment. This extra weight increases the forces acting on the slope, so the failure is caused by a change in both the forces and the internal strength.

Bedding planes or other structures in the slope can provide a plane of weakness along which a failure might be focused. In these circumstances, a slide or flow may occur in material that would have remained stable but for the presence of a structural weakness that can be attacked by percolating moisture. Earthquakes can also trigger mass movements. The vibration applied to slope materials can reduce the internal strength of the sediments by moving the particles and reducing the cohesion or friction between them.

4 Types of Mass Movement

As the classifications above show, there are many different types of mass movement that occur on a variety of scales and in a variety of locations.

a) Creep

Creep is the slowest type of mass movement. Its action can be measured in centimetres by studying the downslope migration of objects such as fenceposts. Often referred to as **soil creep**, this slow method of operation is highly effective. It has been suggested that more material ends up at the bottom of a slope as a result of the action of creep that by any other mass movement process. In temperate environments, soil can creep downhill at between 1 and 2 mm per year, while tropical environments experience slightly faster rates of 3–6 mm per year, and semi-arid areas with cold winters might experience rates of up to 10 mm per year.

There are a number of processes acting on soil that cause it to creep:

- The expansion and contraction of water within the soil is the main cause of soil creep. Seasonal **wetting and drying** leads to the contraction of the soil during the dry season and its subsequent downhill expansion under gravity when it is next wetted. Diurnal and seasonal **freeze–thaw** cycles will also cause downslope movement. Frost heaving within the soil will force particles upwards. When the ice melts the particles will then drop downhill under the influence of gravity. It is this process that explains the high rates of soil creep in semi-arid regions with cold winters.
- **Rainwash** and **rainsplash** (the impact of raindrops) may move small particles downslope. Goudie (1993) also suggests that rainsplash on the downslope side of large stones can remove finer material that was supporting the stone, thus facilitating the movement of larger particles as well.
- Finally, the action of animals can encourage soil creep. As animals, both wild and domesticated, follow tracks across slopes the forces they apply to the soil or weathered material will encourage that material to move downslope.

There are some important variations on soil creep that have great significance in some parts of the world. **Solifluction** is the downhill movement of soil that has become saturated. It is common in periglacial environments where the opportunities for saturation are greater because of the impermeable nature of the permafrost layer. The process of solifluction, when it occurs in periglacial areas, is a combination of two processes, demonstrating the difficulties in using classification systems. There is creep caused by the freeze–thaw action within the saturated layer and flow caused by the movement of water-soaked debris overcoming the internal resistance to gravity. Solifluction produces some of the most obvious landscapes associated with creep. **Solifluction lobes** are rounded domes of soil or rocks that have been moved downslope by the process. They are common in periglacial areas, such as Alaska and Siberia. Often the lobes have flat tops known as **solifluction terraces**. Generally, solifluction in periglacial areas creates an uneven and hummocky terrain where the layers of rock and soil are very disturbed and distorted, and vegetation is often disrupted as well. Some authors differentiate between solifluction and **gelifluction**. Gelifluction is a term introduced in the 1950s to describe solifluction that is associated with frozen ground where the dominant process is that of frost heave rather than flow of saturated soil.

b) Flow

As Carson and Kirby's classification suggests, **flows** take place when the slope material is at its wettest. This means that these mass movement processes are continuous and involve a discrete layer of the slope, as shown in Figure 5.3(a).

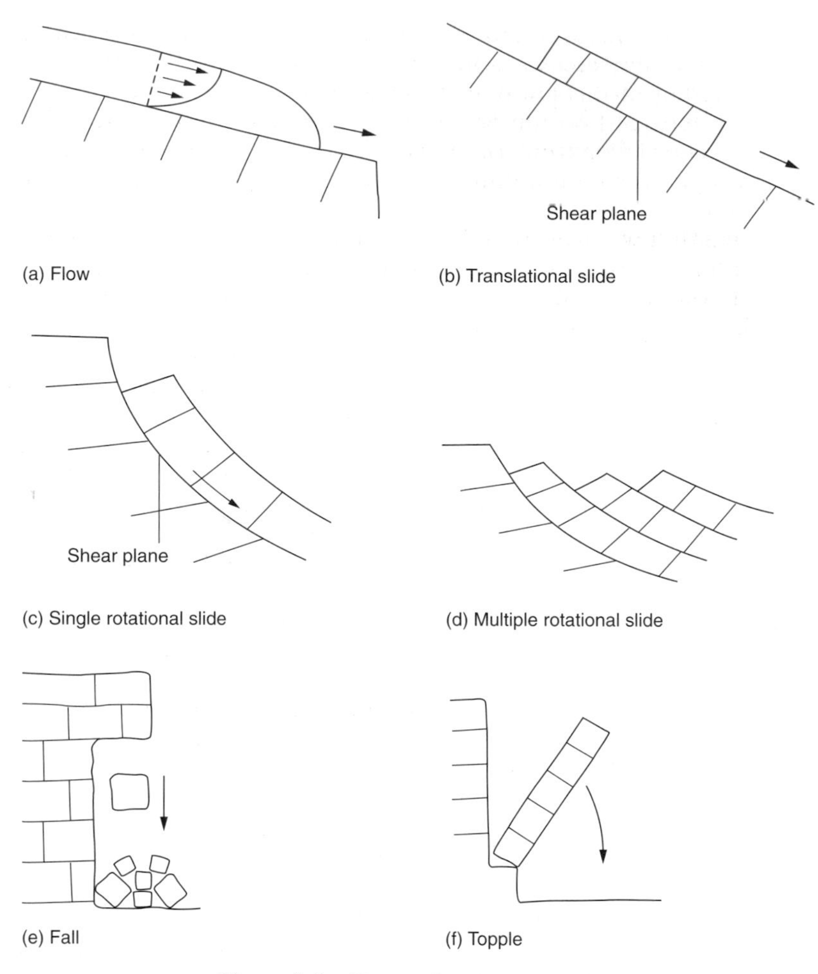

Figure 5.3 Types of mass movement.

Flows occur in three main forms, **mudflows, earthflows** and **debris flows**. The scale of sediment involved increases, ranging from clays to blocks of rock or boulders. Water content varies as well, flows in clay will occur when it is saturated, as suggested in the discussion of solifluction above. These three different sorts of flow are worth considering separately:

- A **mudflow** occurs when fine textured sediments such as clays or some soils become saturated and flow downhill because the excess water has reduced their internal strength. They may contain up to 60% water by weight and move very rapidly, with the potential to make significant alterations to the landscape, or to cause a major disaster to a human community. Volcanic eruptions can cause huge mudflows known as **lahars** when snowmelt added to debris loosened by the eruption flows down the mountain at high speeds. The eruption of Mount St Helens in 1982 produced a lahar that did far more damage than the lava flows. The same can be said of the eruption of Mount Pinatubo on the Philippine island of Luzon in 1991.

- An **earthflow** involves soils and other loose sediments sliding downhill in a viscous mass at a much slower pace than a mudflow. Earthflows are usually confined to a channel of some sort, and lose their velocity quickly when they reach a flat surface. Thus, earthflows often form distinctive lobes when they arrive on a valley floor. The slower pace of earthflows makes them less dangerous to people than mudflows, but they can still do damage to infrastructure such as roads and railways. Concentration of rainfall on one area of the ground as a result of the position of vegetation can trigger an earthflow, and in areas of high rainfall it is not uncommon for roads to be blocked by earthflows from neighbouring woodland, especially if the slope has been disrupted by the construction of that road.
- **Debris flows** involve the rapid movement of relatively coarse material, and are sometimes known as **debris avalanches**. They commonly occur in mountainous areas where the surface is often made of debris unconsolidated by any layers of vegetation. Heavy rain can mobilise this material and turn it into a slurry that resembles cement. Flow tends to follow existing channels, and move at between 3 and $10\,m\,s^{-1}$, carrying large boulders amongst smaller material. If they come into contact with human activity they can do significant damage. Cars have been carried away by debris flows, and roads and railways destroyed. Culverts and bridges can be blocked, causing an increased flood risk, especially since heavy rainfall is needed to create the debris flow in the first place. Figure 5.4 shows a mountain road near Imlil in the Moroccan Atlas Mountains that has been breached by a small debris flow. The channel that the debris flow has followed is visible in the background. The previous 48 hours had contained abnormally high levels of rainfall.

c) Slides

Slides differ from flows because they are a dry mass movement. Hence, they occur in the opposite corner of Carson and Kirby's flow classification. Because they do not involve the reduction of the internal strength of the slope material created by friction of cohesion, they rely on a weakness already present in the slope as their causative mechanism. The term landslide has entered everyday vocabulary, which can provide some problems for definition, since when used in everyday speech the term encompasses flows and falls as well as slides.

Figure 5.4 Debris flow near Imlil, Morocco.

In a pure slide, failure occurs along a well-defined **shear plane**. A shear plane is a plane within the slope along which the failure occurs. **Shearing** is the process that occurs when the two parts of the slope slide past each other. Shear planes are marked on Figure 5.3(b, c), which shows the two main types of slide.

- A **translational slide** is a slide where the shear plane is parallel to the surface of the slope, as shown in Figure 5.3(b). Three characteristics of translational slides have been identified:
  - Steep slopes and adverse geological structures, such as joints dipping steeply downslope to provide a potential shear plane underneath a steep slope.
  - **Weak layers** supporting heavier rocks. Layers of shale beneath thick strata have played a prominent part in several landslides. Wet shale lacks the strength of other rocks and can provide a layer for heavier rocks above to slide over.
  - A **trigger mechanism** is needed. Slides are initiated by either a natural event such as an earthquake or by human action such as oversteepening or undercutting a slope.

  Not all translational slides have a shear plane that represents a change in geology. In 1959 an earthquake in Madison County, Montana, USA triggered a landslide that dammed the Madison river completely. On a steep slope above the river, steeply dipping layers of limestone were adjacent to a thick band of heavily weathered metamorphic rocks (gneisses and schists). The earthquake shook the limestone until blocks broke free high up on the slope and plunged into the valley. Once the limestone was weakened the gneisses and schists followed suit and the valley was soon full of sediment. Fortunately, in this case there was no loss of life, but the Vaiont dam disaster, discussed in a case study at the end of this chapter, was an altogether different story.

- **Rotational slides** or **slumps** are shown in Figure 5.3(c) and have a curved concave shear plane. The rotational movement within a rotational slide moves material from the higher parts of the slope to the lower, and that sediment that was at the lower end of the slope is pushed outwards, often forming a lobe protruding into a valley. Rotational slides can have multiple layers, as shown in Figure 5.3(d), examples of which can often be seen on soft cliffs such as those at Overstrand on the Norfolk coast. Coastal slides are not uncommon, especially on unconsolidated cliffs such as the boulder clay of Norfolk. Saturation of the base of the cliff by waves causes the slide to begin, leaving a steep scar, often crescent shaped, at the top of the slide. Much larger lumps were created in the Turnagain region of Anchorage in the 1964 earthquake. The city is built on an 18 m thick layer of gravel on top of clay. As the ground shook the saturated clay liquefied, and the stable gravel slid over the suddenly mobile ground. As high ground slid downwards rotational slides occurred throughout the city, but most notably in the then fashionable Turnagain Heights area that overlooked the Cook Inlet. The steep gradient between the residential area and the sea was ideal for rotational slides to develop and the housing was completely destroyed.

d) Falls

Falls are generated on the steepest slopes, allowing debris to fall through the air before accumulating at the base of the slope or cliff. Falls will occur when the weathering processes acting on a cliff face have done sufficient damage to allow a block of rock to become detached and fall under gravity. Conical accumulations of debris at the base of cliffs or steep slopes are known as **talus cones**. The presence of a talus cone is an indication that falls are an active process on a given cliff, and should act as a warning to people to be careful. A vegetated talus cone suggests that falls were once active in that location, but are no longer, since the presence of vegetation shows that the cone has been undisturbed for long enough to allow vegetation to begin to grow. Falls can occur as falls (Figure 5.3e) where the rocks drop vertically off the cliff or **topples** (Figure 5.3f) where the rocks fall forward, pivoting around a hinge point.

e) Subsidence

Like falls, **subsidence** is a vertical form of mass movement, as described in Varne's classification. It is the lowering of the surface following the removal of some part of the structure that previously supported it. Of all the mass movements discussed above it is the one perhaps most likely to be caused by human activity such as mining and these causes will be discussed further in Chapter 7.

Away from human causes, subsidence can occur in periglacial areas where permanently frozen bedrock is interspersed with unfrozen material. This situation, known as **discontinuous** or **sporadic** permafrost, can lead to the creation of uneven terrain where the ground surface is lower in the unfrozen areas because of the smaller volume of the water there. This is a landscape known as **thermokarst**. In limestone areas the creation of swallow holes and dolines can lead to the subsidence of the surface material as it descends to fill in the gaps in the expanding joints.

5 Human Activity and Mass Movement Processes

a) Causes of human-induced mass movements

As we have seen, many factors can cause mass movement processes to begin, but there is no doubt that one of the most effective agents is human activity. Goudie and Viles (1997) put it succinctly when they said "human capacity to change a hillside and to make it more prone to failure has been transformed by engineering development". In the days before the existence of mechanised diggers and the need for improved infrastructure, people existed in some degree of harmony with the topography around them. Certainly, they were unable to make significant alterations to that topography because powerful tools were not available until the Industrial Revolution had provided the means. Now, however, the situation is very different. Excavations are deeper, and taller buildings are being built, roads and railways are following increasingly convoluted routes, and land that was not used for construction before is now being utilised because of the pressures from increasing populations. Land that is vulnerable to slope failure can be classified as **marginal land**, a term that includes all land that is only just viable for local purposes.

There are a number of ways in which human activity can inadvertently trigger a mass movement:

- By removing underlying or lateral support through construction of cuttings or excavations, or through draining lakes or reservoirs.

- By building up pressure through the accumulation of buildings or stockpiles of ore or slag.
- By altering the water table through the construction of a reservoir (see Vaiont Dam case study on page 93).

b) Undercutting

Perhaps the most common source of human-induced mass movement events is undercutting a previously stable slope. This is quite a common practice in the construction of **roadways**, especially in mountainous terrain, but it can have unpleasant consequences. Landslides are common events in the Himalaya, especially during the high intensity rainfall of the monsoon season. Slopes become saturated, there are many faults that could act as shear planes, and there are an increasing number of roads and settlements built along the foot of slopes overlooking river valleys. On 12 September 1995 near Kulu in Himachal Pradesh, India, a landslide occurred that killed 65 people. An estimated 96 000 m^3 of unconsolidated sediment slid down the final section of an enormous alluvial fan that rests on the side of the valley of the river Beas. Heavy rains between 3 and 6 September had caused the river to reach bank level, flood a roadway that had been cut into the foot of the slope and begin to erode the unconsolidated foot of the slope. The heavy rain had also caused some saturation within the slope. The road was cut into the foot of the fan, meaning that the final, gentle part of the slope where it meets the flat floodplain of the river valley had been removed. Without the road, the slope would have been in a more stable condition and less vulnerable to further undercutting by the fast flowing and turbulent river. The flash flood caused by the displacement of water as the material from the slope entered the river was the main cause of death as it hit towns and villages further downstream.

Problems associated with road construction are not limited to tropical or mountainous regions. When the Sevenoaks bypass was being constructed in Kent in 1966 earth movers cut through two grass covered lobes of material that turned out to be relict landslides. Unfortunately, the disturbance reactivated the processes that had lain dormant for an estimated 10 000 years and the mass movement lobes began to move again, eventually forcing the bypass to be rerouted, at some considerable expense.

c) Construction

Clifftop buildings are another method by which people have inadvertently exacerbated mass movement processes. The addition of a **heavy object** to the top of a cliff will increase the gravitational force being applied to that slope, increasing the instability of the cliff and the risk of slope failure. The Holbeck Hall Hotel in Scarborough was the cause and the victim in just such a case. A rotational slump occurred in the back garden of the hotel on 5 June 1993. The slump occurred following a number of drier than average summers, and a particularly wet spring in 1993. The top layers of the cliff are made of **boulder clay**, unconsolidated material deposited by retreating glaciers. The unfamiliar rainfall saturated the boulder clay that had become dry and cracked over the previous few years, making it more vulnerable because the water could make its way through the cracks quicker that would otherwise have been the case. As the foot of the cliff undergoes attack by the sea, it is therefore not as stable as a slope with an accumulation of debris at its foot. Although the causes of this slump are at least threefold (saturation, undercutting of the cliff by coastal erosion and the extra weight of the hotel) there seems little doubt that the presence of the hotel was a significant contributory factor in causing the flow.

Cities such as Hong Kong, Rio de Janeiro and Kuala Lumpur are expanding onto marginal land in order to accommodate the demands created by increasing **population pressures**. In doing so there is an increased risk of a mass movement disaster such as a landslide or debris flow. In Hong Kong landslides have killed nearly 500 people since 1948 as new construction has taken place in steep land on the edge of the ever-expanding city. New roads and buildings have been built on steep slopes that have been deforested in order to provide the room for the buildings. The deforestation reduces the stability of the slope because of the removal of the binding properties of the roots, and also exposes the soil to the direct impact of the torrential downpours of the monsoon. Since 1977 the Geotechnical Engineering Office has been responsible for monitoring new developments in Hong Kong, paying specific attention to building new roads and buildings in a way that will do least damage to the slope and therefore minimise the risk of landslides. They also advise on **land-use plans** to minimise public risk, and have instituted a landslide early warning scheme in conjunction with the emergency services.

CASE STUDY: NEVADO DEL RUIZ, COLUMBIA

Nevado del Ruiz is the highest active volcano in Columbia. Its summit is 5389 m above sea level and it is covered in 25 km^2 of snow and ice. On 13 November 1985 an explosive eruption from the mountain's summit crater began and the broad, conical summit was gradually covered in pyroclastic flows and volcanic material. Soon ash and pumice began to fall to the north-east, along with the heavy rain and snow that had begun earlier in the day. The summit area was quickly covered in deposits of ash and other volcanic material to a depth of some 8 m. The hot material of the pyroclastic flows quickly mixed with the snow and ice covering the summit of the mountain, melting the ice and being carried away by the newly formed streams. Soon channels were being cut into the ice caps, up to 100 m wide and 4 m deep in places. The channels contained a mixture of ice, water, pumice, ash and other rock debris, and the contents soon flowed down the mountain to join the main rivers draining Nevado del Ruiz. These flows happened as a series of small lahars swept into the channels of the Azufrado and Guali rivers in particular, but all six major rivers emanating from the mountain were affected to some degree.

Lahars are a much more powerful erosive tool than river water, and as the lahars swept down the Guali Valley they eroded soil, loose rock debris and picked up vegetation from the channel and surrounding area. The river channels were not big enough to cope with the sudden increase in volume of flow and so the floodplains of the rivers were soon covered by the fast flowing (up to 60 km h^{-1}) mudflow. The additional material that was picked up by the lahar meant that the size of the mudflow increased by up to four times the initial volumes, meaning that in some of the narrow canyons further down the mountain the lahars were as much as 50 m deep.

The only safe locations in the upper valleys of the rivers such as the Guali were high up on the valley sides. Eyewitnesses reported that the lahars flowed in pulses; with peaks of particularly high discharge separated by lower levels of flow. This may have been caused by the occurrence of further eruptions that melted more snow and ice and added additional impetus to the flows.

Within 4 hours of the beginning of the eruption, the mudflows had travelled 100 km and left a swathe of destruction in their wake. The cold statistics are:

- more that 23 000 people killed
- about 5000 people injured
- more than 5000 homes destroyed.

Hardest hit was the town of Armero, at the mouth of the Rio Lagunillas canyon. The town no longer exists and three-quarters of its 28 700 inhabitants died in a lahar that covered the town in a layer of mud more than 8 m thick. The first pulse of mud hit Armero at 11.25pm and consisted of a flood of cold, relatively clean, water that overflowed the usual channel of the Rio Lagunillas. This water had been displaced from a lake just upstream when that lake was first hit by the lahar, and was travelling in front of the lahar rather as a bow wave travels in front of a boat. Ten minutes later a second pulse arrived. This was much larger and destroyed most of the town and killed most of the inhabitants in a period of between 10 and 20 minutes, leaving a layer of mud between 2 and 5 m deep across much of the settlement. The third pulse, at 11.50 pm, was only half the force of the main wave, but deposited another few metres of mud across the town. Finally, in the next hour or so a series of smaller pulses swept into the town, experienced only by the few survivors.

The destruction of Armero in 1985 was the third time in human history that a settlement in that location had been destroyed. In 1595, shortly after the arrival of the Spanish colonists, and again in 1845, hundreds of people died when mudflows from Nevado del Ruiz swamped this location. Each time people forgot the previous tragedy and the site, which is on a debris fan close to the Rio Lagunillas, has been reoccupied. It is an attractive site at first glance, close to the water source, but high enough above it to be away from the risk of normal flooding.

Figure 5.5 shows the site of the town of Armero today. Some houses that were on the outskirts of the town can be seen on the far left of the photograph.

The disaster on Nevado del Ruiz can provide a good deal of useful information to those studying lahars, in the hope that other towns vulnerable to a disaster such as that which destroyed Armero might be better prepared for such an event:

- A relatively small eruption on a snow and ice covered summit is enough to create a devastating mudflow.
- A lahar will only occur if there is a mechanism for mixing the hot volcanic debris with the snow and ice. In the case of Nevado del Ruiz the falling heavy rain and snow provided that mechanism.

Figure 5.5 The site of Armero today.

- Lahars are likely to grow once they have started to flow and will increase their volume significantly if they flow over a landscape that has large supplies of already eroded sediment for them to entrain.
- A lahar that is confined in a valley is likely to maintain high velocities until it reaches a location where it will spread out laterally and dissipate its energy. Armero represents one of these sites, as it sits at the mouth of the gorge. Mudflows will therefore concentrate there.

CASE STUDY: THE VAIONT DAM DISASTER OF 1963

One of the most damaging landslides of modern times happened in 1963 at Vaiont in northern Italy.

The river Vaiont, a tributary of the Piave, had been dammed for 3 years a kilometre or so from the confluence with the Piave. The Vaiont flows out of the Italian Alps into the Piave, which makes its way south to the Adriatic. The site of the dam and reservoir is a narrow, glaciated valley flowing steeply out of the mountains. The area immediately to the west of the valley is the populated lowland valley of the Piave river. During the 1950s, Italy was repairing and rebuilding after the second world war, and emphasis was placed on using alternative sources of energy where possible. The Vaiont Valley was identified as a location with some potential as a hydroelectric installation. There was a market nearby, and the topography of the valley was ideal for the construction of a short but high dam to create a reservoir in the narrow glaciated valley.

The Italians built what was at the time the second largest dam in the world. The crest of the 190 m long dam was 262 m above the river and it held back a reservoir of 150 million m^3 of water, the third largest reservoir in the world.

Despite the magnificence of the engineering achievement in creating such a dam, the benefits of the project were short lived. Evidence that not all was well first came in 1960 when 700 000 m^3 of rock slid into the reservoir as it was being filled with water. Measurements were also being taken of the creep rates on the south side of the reservoir, and, by January 1963, they had reached about 1 cm a week well above normal creep rates. As the level of the reservoir rose, so did the creep rates, and by the time the lake was full, in September 1963, the rate of creep had reached 40 cm a day.

On 28 September 1963 a period of heavy rain began, and on 1 October a shepherd reported that his sheep would not graze the land on the north slope of Mount Toc, which is to the south of the Vaiont reservoir. By 7 October engineers were concerned at the amount of water that was running directly into the reservoir through overland flow, and at the still rapidly increasing creep rates. They opened outlets in the dam to try and reduce the water levels and alleviate the stresses being put on the rocks around the dam. However, their actions came too late.

On 8 October 300 million m^3 of rock and soil slid down the north slope of Mount Toc at more than 90 km h^{-1} and plunged into the Vaiont reservoir. An area of mountainside nearly 600 m high, 150 m deep and 2.4 km wide had collapsed and filled the reservoir with debris 150 m above its former level.

The movement of the rock and soil sent a blast of air, water and rock soaring into the air with such force that the roof was taken off a house 250 m above the reservoir. Displacement of the reservoir's water meant that a wall of water 100 m high was sent over the Vaiont dam and down to the valley of the Piave below. The town of Longarone, near the confluence of the Vaiont and the Piave, was engulfed by a 70 m deep wave which then spread up and downstream for many kilometres, flooding both sides of the populated valley. In the 7 minutes it took for the Piave Valley to flood nearly 3000 people lost their lives. Bizarrely, the dam itself sustained no noticeable damage, but the reservoir that had been going to power the villages and towns of this area of northern Italy was full of rock and soil.

Causes of the Vaiont disaster

The tragedy that occurred in 1963 was caused by insufficient understanding of the geology of the area and the likely impacts the construction of a dam and a reservoir would have. Figure 5.6 is a geological cross-section of the Vaiont Valley. It is a fault-controlled valley cut into limestone, shale and clay beds. On the north side the limestone and clay dips steeply towards the valley floor. All the rocks in the valley had been weakened by weathering, especially freeze–thaw action, since the valley is situated in the foothills of the Italian Alps.

In addition to the unfavourable geological setting, the filling of the reservoir caused a rise in the levels of the groundwater in the Vaiont region. The water penetrated the bedding planes and joints in the limestone and saturated the clays, reducing the cohesion of the rocks and weakening the structure along several potential shear planes. The heavy rains in the week preceding the disaster merely served to accentuate the problems, which had been identified through the monitoring of the creep rates on Mount Toc.

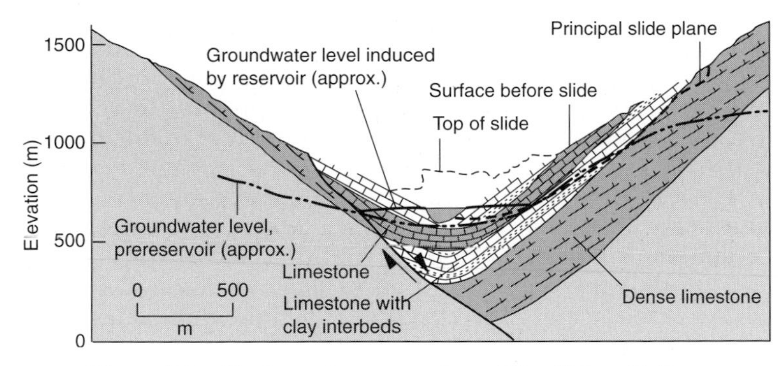

Figure 5.6 Geological cross-section of the Vaiont Valley.
Source: Dolgoff (1996).

Figure 5.6 shows how the groundwater level rose as the reservoir filled, from the floor of the valley before the construction of the dam to the surface of the reservoir once it was full. This rise in the groundwater destabilised the lower slopes of Mount Toc, especially along the clay beds within the limestone. Once the slide began there was nothing to support the higher slopes and they collapsed as well, along the line marked as the principal slide plane. The diagram also shows the new land surface. This filled in the valley along the length of the reservoir leaving a tiny lake adjacent to the dam as the only remaining reservoir.

The Italian judicial system eventually prosecuted those responsible for the construction of the Vaiont dam, but the episode remains as a salutary reminder of the importance of understanding that human activity has an impact on the natural environment, which is not always able to adjust.

CASE STUDY: THE NICOLET LANDSLIDE

A relatively unusual structural arrangement of deposits left by retreating ice sheets has caused a number of earthflows in Norway, Sweden and parts of Canada.

When glacial action has deposited layers of clays, sand and silts on outwash plains to create flat-topped terraces adjacent to rivers and lakes an environment has been created that seems prone to earthflows. On several occasions, the upper layers of these arrangements have slid, apparently spontaneously, into the river. The St Lawrence river has experienced a number of these landslides. The upper layers are made of silts, and below them is a layer of soft clay. It is this layer of soft clay that acts as a shear plane and allows the silt to move. In 1955 one of these earthflows, which occur on very shallow slopes, occurred in the middle of the Canadian town of Nicolet, Quebec, carrying much of the town into the Nicolet river. Fortunately, there were only three deaths as a consequence, but much of the infrastructure of the town was destroyed including a critical bridge.

This earthflow occurred because the layer of soft clay is in fact **quick clay**. Quick clay will turn from a solid state to a near-liquid condition if it is subject to a shock or disturbance. Layers of quick clay are thought to have formed in shallow waters of saltwater bays near to the end of the Ice Age. When the thin plates of clay are laid down they adopt a 'house of cards' structure, with large water-filled pore spaces between them. The chemical bonds made by the saltwater hold the plates of clay together and give the layer some solidity. However, when these layers are elevated above sea level, as often happens as a result of the crust rebounding after being depressed by the mass of the ice sheets, this is known as **isostatic readjustment.** After the melting of the ice sheets, the salt water is replaced by freshwater that does not have the same binding qualities. A mechanical shock, such as a small earthquake, can cause the structure within the clay to collapse and then the large quantity of water within the layer will turn it into a very wet clay mixture that behaves almost as a liquid when it is on a slope. As a result the layers above the quick clay are left with almost no support and earthflows can begin. The magnitude of the earthquake necessary to trigger this process is not great, indeed a tremor that could not be detected by human senses would be sufficient to liquefy the clay.

Summary Diagram

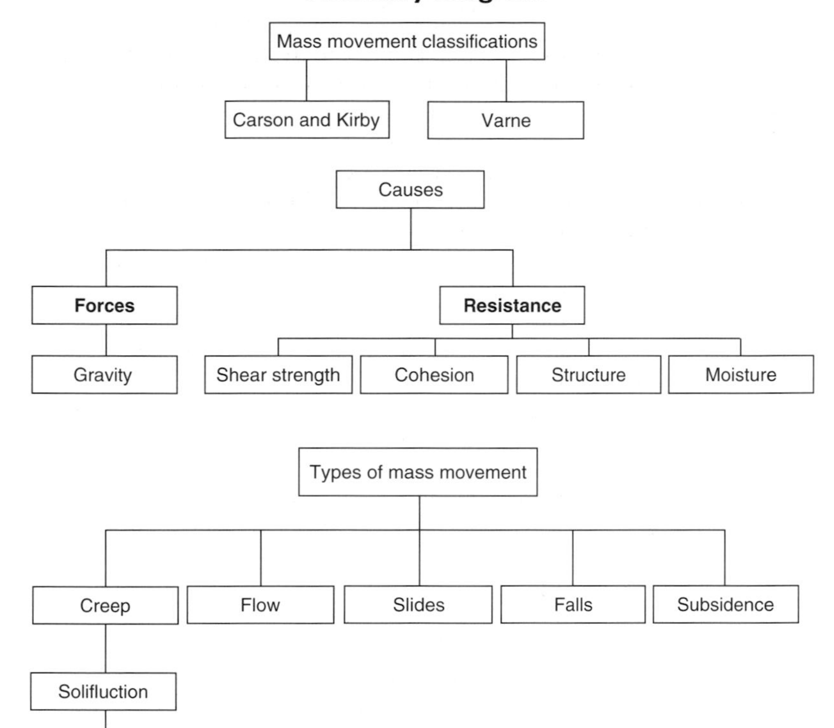

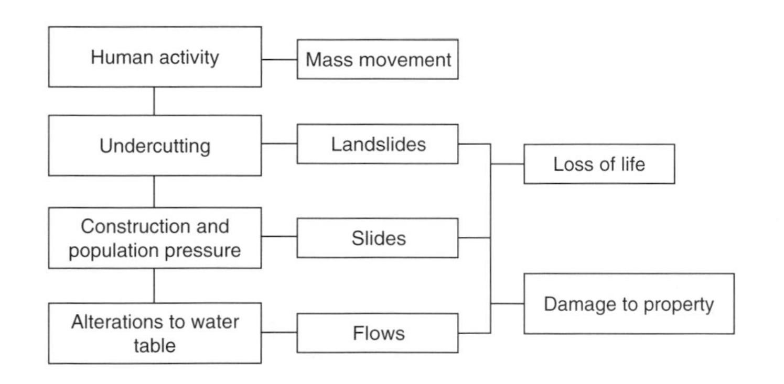

Questions

1. What are the causes of mass movements?
2. Describe and explain the relationship between mass movements and
   a) rock type
   b) gradient
   c) vegetation.
3. What is the impact of human activity on mass movement?
4. What is solifluction? Why is it an important process?
5. Draw annotated diagrams to illustrate and explain each type of mass movement.

6 Slopes

KEY WORDS

Equilibrium: a state of stability or balance, where the nature of the slope remains the same because the inputs balance the outputs.
Slope profile: the vertical structure of a slope along a given cross-sectional line.
Slope unit: the individual elements within a slope, that combine to create the full slope.
Slope evolution: the way a slope changes over time. Three hypotheses have been put forward: slope decline, slope retreat and slope replacement.

1 The Slope System

There are slopes everywhere, and they dictate and direct many human activities such as agriculture, road and railway construction, building and drainage. Many of the processes that act on them have been discussed in the previous chapter; mass movement, and the case studies covered there, are equally applicable to this chapter.

Many slopes seem to be extremely simple in form and show no obvious signs of change or development. However, they vary greatly from one location to another and have many characteristics that can be studied. Every slope represents the interaction of weathering processes, transport processes, rock type and climate, and geomorphologists have long been engaged in the task of producing models to show how slopes evolve over time.

Slopes are, at a simple level, the result of active processes shaping passive materials with the slope form being the end product. However, at a more complex and realistic level, slopes are created as the result of numerous linkages between **factors**, **processes** and **forms**, as shown in Figure 6.1.

The systems approach to geographical study focuses on **inputs** and **outputs** as a means of understanding the creation of a landform. In the case of a slope, the inputs can be broken into two categories:

- Inputs of **energy**. Energy comes from incoming solar radiation, falling raindrops and wind. Also, all slopes have a stored reserve of gravitational energy that influences the downward movement of material, as discussed in the previous chapter.

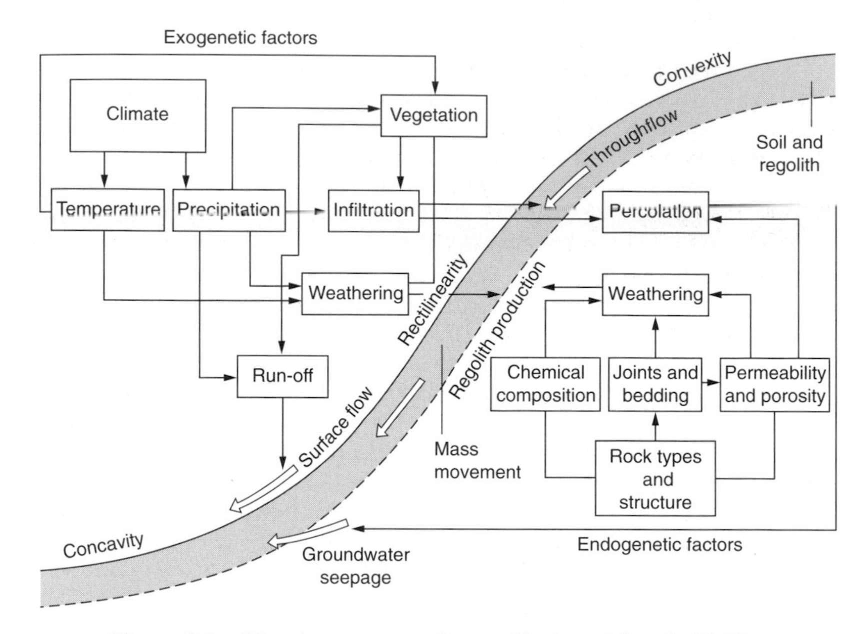

Figure 6.1 The slope system. *Source*: Clark and Small (1982).

- Inputs of **mass**. Material is added to slopes in many forms. Water comes from rainfall, snowmelt, springs and seepages. Inorganic matter comes from the weathering of bedrock, through the processes outlined in Chapter 4, and organic matter is provided by flora and fauna.

To balance these inputs there is a set of outputs, also dealing with energy and mass:

- Energy is lost through the removal of heat through nocturnal radiation or through movement of the slope material.
- Mass is lost as streams remove water, debris, solutes and organic material, and mass movement processes will do the same job if conditions permit.

The energy and the material can also be **stored** on the slope, completing the system. Slopes are an **open system**, because all the inputs to a slope system come from another location, and as all the outputs leave the slope to go to another location it is not possible to study slopes in isolation. Climatic conditions dictate the quantity of rainfall and heat energy. Consequently, climatic change might have an impact on the behaviour of a slope system. Fluvial systems and processes are closely linked to slopes because much material is removed from slopes by water. Equally, erosion by streams may cause the **steepening** of slopes due to undercutting or downcutting. This will also change the rate of movement on the slope.

The balance between inputs and outputs on a slope might reach a **steady-state** or **equilibrium**. Equilibrium is difficult to identify in the field, because it is hard to establish whether a slope is in equilibrium over a period of time. It may take a slope many years to adapt to a change in its inputs or outputs, especially long-term changes such as climatic change or tectonic uplift. As these changes occur, then the slope will respond, and this response may take the form of occasional dramatic events, or minute movements occurring frequently over a long period of time.

For example, a change in climate from humid to semi-arid will lead to a reduction in the amount of vegetation cover on the slope. This will increase the efficacy of running water as an agent of erosion and transport and the vulnerability of the surface material to removal. This increase in removal of material is likely to cause the reduction of the slope angle as debris is removed from the upper part of the slope and redeposited at the base.

Using this systems model of slope analysis enables a common pattern to be applied to slopes in different climatic or geological conditions, and to understand the impact that different types of human activity might have on a slope.

2 Slope Forms

The shape of any slope is the product of the balance of inputs and outputs, but each slope can be broken down into a sequence of component parts. There are two main ways of dividing slopes up into these parts.

a) Slope function segments

Three main types of slope unit exist within this category.

- First, **denudation slopes** are those that are experiencing a loss of material. Denudation slopes can be subdivided into those that are controlled by **weathering** and those that are controlled by **erosion**. In a **weathering-controlled denudation slope** the rate at which weathered material is produced is less than the rate at which it can be removed. Thus, the amount of material removed from such a slope is dependent on the rate of weathering because weathered material is removed as soon as it is produced. These slopes tend to have very little loose debris on them. On an **erosion-controlled denudation slope** the rate of removal of material is slower that the rate of production of removable material. These slopes are often covered in thick layers of boulders and debris.
- Second, **transportation slopes** are those that do not experience loss or gain of material because the amount of material added to the top of the slope is matched by the quantity removed at the bottom.

- Finally, **accumulation slopes** are those on which the amount of debris is increasing.

These terms should only really be applied to a point on a slope, but slopes can often be broadly divided according to these terms. A weathering-controlled cliff (denudation slope) might have a steep transport slope below it, which in turn will have a talus cone or scree slope (accumulation slope) at its base. The cover photograph shows areas that are clearly erosion-controlled slopes and others that are weathering-controlled slopes. The summit regions of the mountain show no loose debris and are weathering controlled, while the slopes below the saddle on the left-hand side of the picture have accumulations of debris indicating that they are erosion controlled.

b) Slope angle segments

The angle of a slope allows that slope or section of slope to be categorised according to that angle. The slope angle segments are put together to create the **slope profile**. An idealised slope profile showing the segments described below is shown in Figure 6.2. As with slope functions, there are three main categories.

- A **straight** slope is a slope on which the angle does not change with distance. A subsection of this category is the **cliff**, which is a vertical straight slope. A straight slope can be created in a number of ways. Very active undercutting or weathering at the base of a slope may cause a slope failure. After the failure the slope will rest at an angle that is related to the strength of the material and its moisture content. Straight slopes are commonest when rapid types of mass

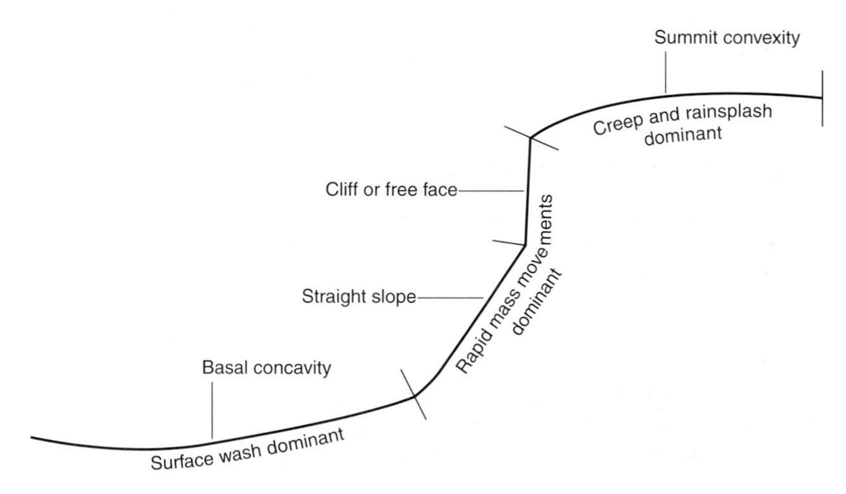

Figure 6.2 An idealised slope profile showing the four main slope segments. *Source*: Goudie (1993).

movement are a dominant process. The straightness of a cliff will be determined by the joints in the bedrock.

- A **convex** slope is one on which the angle increases with distance downhill. They tend to occur on the higher portions of slopes. Slow soil creep processes tend to dominate on such slopes as they curve away from the summit.
- A **concave** slope is one on which the slope angle decreases with distance downhill. Concave slopes tend to occur at the base of a slope profile where the washing of sediment onto and across the slope segment is the dominant process. The reason for the concavity is that energy levels in the streams feeding sediment on to the segment will decrease as soon as the gradient drops, causing more material to be deposited nearer the steeper part of the slope.

This four-segment model was first suggested as long ago as 1942 and has the advantage of simplicity and adaptability. Elaboration of this idea is possible, and a five-unit model was suggested in 1974, by Caine, as being typical of mountain slopes. This added an additional unit by subdividing the lower straight slope to allow for the likely accumulation of talus in mountainous environments. Another elaboration is the nine-unit profile devised by Dalrymple in 1968. Again, this subdivides the four units by, for instance, dividing the convex unit into an **interfluve** (land dividing two drainage basins) and two creep slopes of different angles. It also adds three units at the foot of the slope to include the slopes typical of a river, such as the channel wall and channel bed.

3 Models of Slope Evolution

The way in which slopes change over time is a topic that has provided much debate among geomorphologists. There are three main theories of slope evolution that have been put forward, those of Davis, Penck and King.

a) Davis' theory of slope decline

Slope decline was an idea proposed by the early American geomorphologist W.M. Davis, who developed his ideas for slope evolution in the humid temperate areas of the Appalachians, the ancient range of hills and mountains in the north-eastern part of the USA. Davis' theory suggests that the steepest part of the slope becomes progressively shallower with time, as shown in Figure 6.3. This reduction in steepness is accompanied by the development of convexity in the upper portion of the slope, and concavity in the lower section, as suggested in the slope unit models of the previous section. Davis' model of slope decline was part of his 'cycle of normal erosion', a seminal work in the development of the understanding of natural landscape.

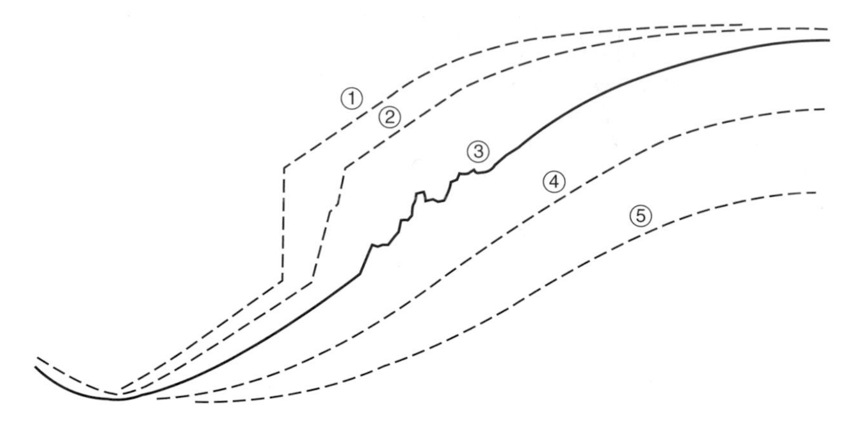

Figure 6.3 A slope decline sequence. *Source*: Clark and Small (1982).

Davis postulated that a slope would first be formed by the rapid downward incision of a stream. As the slope responded to this stimulus he suggested that the balance between transport and weathering would lead to a gradual decline in the angle of the slope as weathered material was transported downhill. This would occur, said Davis, because each point on the slope had to transport the material weathered at that point, and the material that has arrived at that point from further up the slope. As a result the ability of transport processes to clear the slope becomes less with increasing distance downhill because the amount of debris to be removed increases with that distance. Therefore, the rate of weathering, debris removal and ground lowering must be faster upslope, leading to the overall decline of the slope.

Field studies to support Davis' model set out to identify slopes in areas of constant geology and climate that had different dominant angles that could be linked to the age of the landscape. Such evidence was not easy to find, and in many mountainous areas the existence of angular peaks and summits argued against the theory that the upper section of a slope would be lowered first.

Slope decline is most likely to occur in a mature landscape, once a smoothly graded profile has been established.

The final expression of Davis' model of slope decline is a **peneplain**, a gently undulating plain created as vast amounts of material were transported from the upper part of all the slopes in the area to the lower part of those slopes.

b) Penck's theory of slope replacement

Walther Penck was a German geomorphologist who believed that slope evolution was a more complex issue than Davis had suggested.

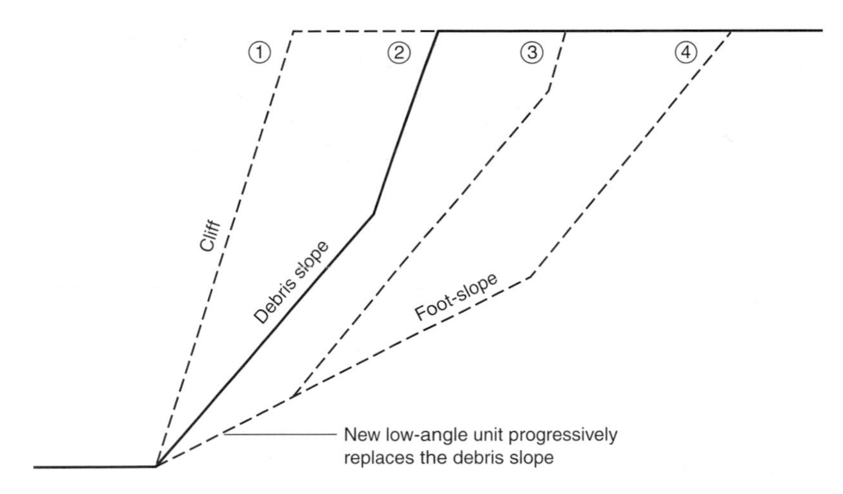

Figure 6.4 Penck's model of slope replacement.
Source: Clark and Small (1982).

Penck worked in much more mountainous areas, such as the Alps and the Andes, where tectonic processes were active. He felt that the impact of uplift was vital to understanding slopes, because that uplift controlled the rate of river erosion, which, in turn, controlled the evolution of the slopes on the valley sides. Related to this idea Penck argued that steep slopes would weather and retreat quicker than a gentle slope, so that a steep slope or cliff is eventually replaced by a gentle slope or debris cone. In turn, this slope will be replaced by a still gentler slope as increasingly weathered material is transported further downslope, as illustrated in Figure 6.4. The end product is a gentler slope profile, but this is not achieved by the 'hinged' decline suggested by Davis. Over time the proportion of the slope occupied by the steeper segments of the profile are reduced, although the maximum elevation of the slope is not reduced. This is another difference between the slope replacement model and the slope decline model.

The final expression of Penck's model is a dissected landscape with wide valleys and gentle slopes rising to isolated peaks.

c) King's model of parallel slope retreat

L.C. King was a South African geomorphologist who developed his ideas for slope evolution by studying the semi-arid landscapes of that area. The dominant feature of these landscapes are steep-sided, island-like often symmetrical hills called **inselbergs** or **kopjes**, shown in Figure 6.5.

King's theory suggests that the vertical part of the slope profile occupies the same proportion of the slope profile as it retreats. The

Figure 6.5 Inselbergs in Monument Valley, USA.

retreat creates an ever-increasing concave unit of accumulation at the foot of the hill. While the angle of the vertical face scarcely changes with time, the concave slope will become increasingly gentle with time. King dubbed this gentle slope a **pediment** and the steep slope a **scarp**. His fieldwork showed that inselberg pediments had angles between 0.25° and 7°, while the scarps had slopes of between 15° and 30°. The example shown in Figure 6.5 is rather more extreme, as a result of geological controls steepening the scarp.

King postulated that the youthful stage of this slope evolution began with the rapid uplift of a previously existing plain. Rivers would then cut rapidly downwards into that plain creating steep-sided valleys. As these valley slopes began to be weathered and eroded the pediment would develop at the slope foot. King then envisaged the slopes retreating parallel to each other, away from the river, leaving ever-expanding pediments and ever-decreasing high plateaux, known as **pediplains**. The end product is steep-sided, isolated hills. A diagrammatic representation of King's model of slope evolution is shown in Figure 6.6.

d) The wider perspective

Although there was a lengthy debate among geomorphologists during the twentieth century about which model was 'right', further research has shown that all models have some validity. In the right

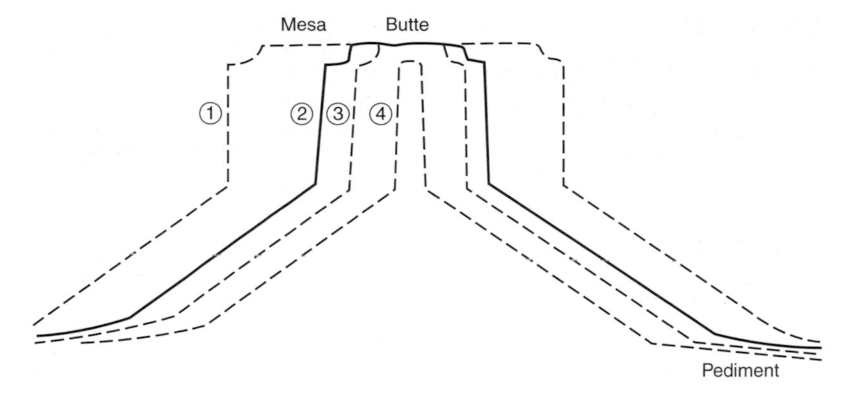

Figure 6.6 A slope retreat sequence. *Source*: Clark and Small (1982).

conditions slopes can be found that fit each model, yet it is unlikely that any such slope reached that situation by following the pattern of model that it then matched.

The variety of factors that influence the evolution of slopes is so great that it is rare to find conditions that match the requirements of any of the models over a large area. There are isolated locations where that has happened, such as the inselberg scenery of Monument Valley.

Different factors can cause changes in a slope's behaviour over time. Tectonic uplift or a change in base level induced by human activity may rejuvenate a stream and cause a change in gradient. Climatic change may alter the amount of water available to the slope for the erosion and transport of material. Each of the three models requires the slope to be subjected to unchanging conditions for long enough for the processes acting on it to reach equilibrium. The mature slope forms outlined in each of the models is a representation of a slope in equilibrium, and that is a state which is not easily achieved in a world where few factors remain constant at a time of increasing human impact on the environment.

4 People and Slopes

There is a clear relationship between people and slopes. Processes of mass movement that take place on slopes can have a direct and damaging effect on human communities, as shown in the last chapter. It is also true to say that much of that effect is caused by human activity. Equally, however, slopes control human activity in a number of ways. **Land use** can be restricted by the nature of a slope, while advanced study of a slope can allow **hazard recognition and avoidance** by increasing understanding of the behaviour of the slope.

a) Land use

Slopes influence the use that human communities can make of the land. The steepness and length of a slope will determine the potential uses and the likely hazards from that slope. **Shallow slopes**, those of less than 5°, have an almost endless list of possible uses, although some functions, such as international airports, need to have almost entirely flat land. Flat or nearly flat land poses greater risk from flooding that steeper terrain, especially if the drainage is poor. This has been particularly in evidence as new housing has been built on flat land near rivers and has then suffered from flooding.

Moderate slopes, those between 5° and 15°, can be used for developments. Houses can be built on these slopes, but industrial land uses, with greater demand for land, usually need flatter land. During the boom in coal mining in the valleys of South Wales in the nineteenth century many houses were built on the steep sides of the valleys as the flat valley floors were used up. Wheeled vehicles, including tractors, cannot operate effectively on slopes steeper than 15°, meaning that moderate slopes can be used for arable agriculture only if farmers are equipped with modern machinery. Contour ploughing helps with this problem. **Steep slopes**, those above 15°, have few intensive land uses. Even road building is difficult, since undercutting of a steep slope can lead to failure. Commercial forestry can be carried out with caterpillar vehicles, but even this becomes unviable above about 35°. The steepest slopes of all have no economic value, although they perform valuable functions as water catchment areas, reservoirs or because of their aesthetic values. However, in tropical environments a combination of population pressures and demand for food have led to the cutting of **terraces** into steep slopes to provide locations where crops, especially rice, can be grown.

b) Hazard recognition and avoidance

The extreme mass movement processes explained in the previous chapter are a significant risk to human life and economic activity. As a result **hazard mapping** is a growing field of geographical activity designed to identify locations likely to suffer from extreme slope processes, and advise planning authorities accordingly. This is a much more complex process than the simple division of slopes into the categories outlined in Section 2 of this chapter, since it involves an understanding of slope and mass movement processes, familiarity with field locations, the use of satellite and aerial photography, and an appreciation of the environmental factors influencing any given location.

The end result of a hazard mapping exercise is usually a colour-coded map showing areas of varying slope stability. The decision about which category of stability to place a particular slope segment into will depend on the factors listed in the last paragraph. These maps can serve as a basis for the introduction of planning laws and regulate land uses in areas deemed to have a particularly high risk of slope failure. Insurance companies use them to identify houses in areas of high risk and adjust premiums accordingly.

Many specifically geological factors are included in hazard analysis:

- The relationship between bedrock structure and topography. If rock strata dip towards an undercut, such as the sea or a river, they are more likely to fail. This risk of failure will be increased if further undercutting is carried out to construct roads or other buildings.
- The condition of the bedrock. Highly fractured rock is more likely to fail, as is a rock structure that has dense layers of rock above strata of low shear strength.
- The composition of the slope. A slope made primarily of bedrock is less likely to fail than one constructed of loose material. Even within loose material some soil constructions will be more stable than others. A densely packed soil will be less likely to fail than a boulder clay. A vegetated slope will be less likely to fail than a bare slope.
- The groundwater conditions. A variable water table may promote slope failure, as might the human-induced raising of the water table (see the Vaiont dam case study on page 93).

Summary Diagram

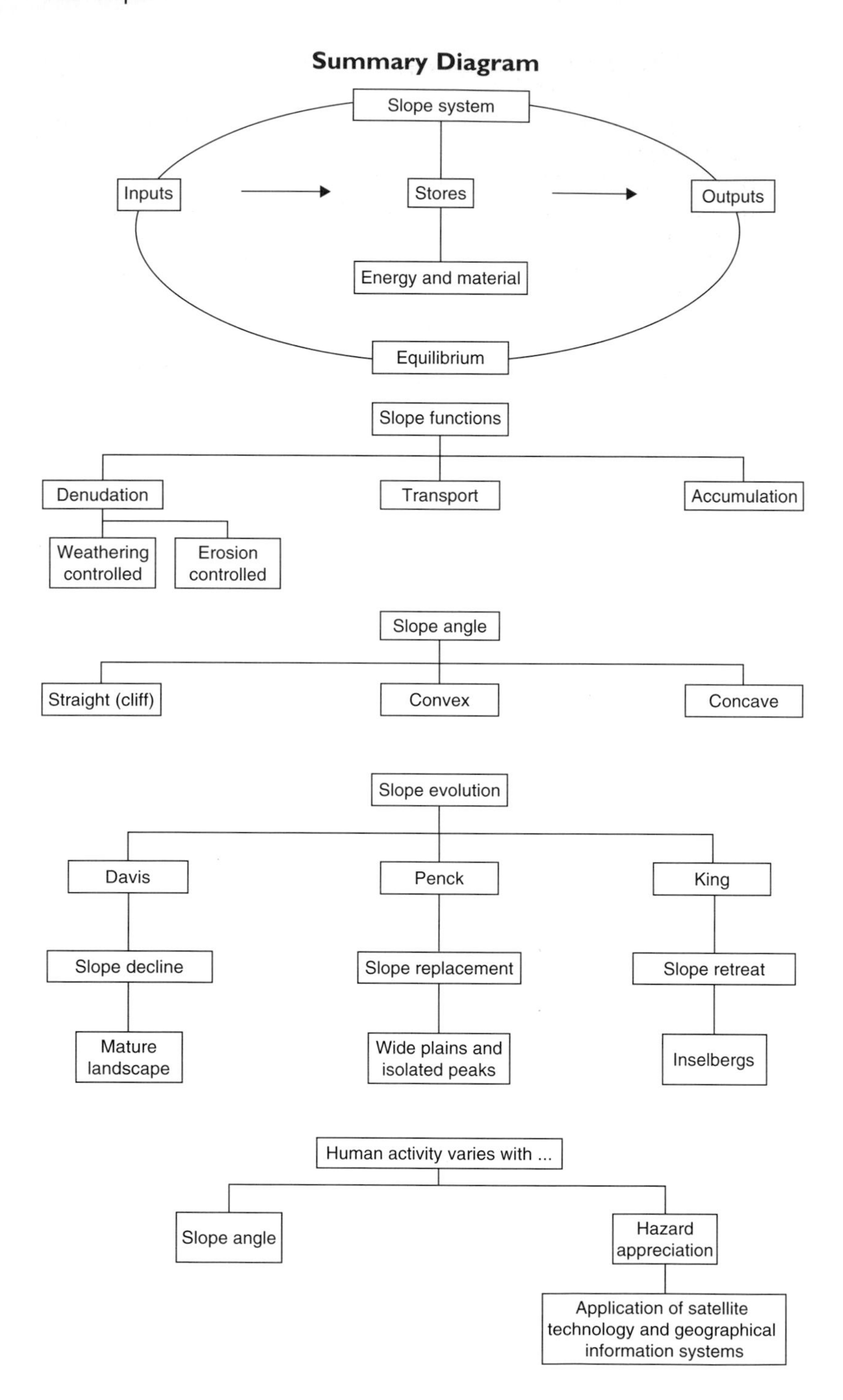

Questions

1. Outline two ways in which **a)** climate and **b)** geology affect slope development.

2. **a)** Name two possible inputs into a slope system.
b) Name and explain two processes of mass movement that can operate on a slope.
c) Describe and explain one way in which a river could influence the steepness of slopes.

3. Examine the relationship between vegetation, slope angle and slope stability.

4. Explain how it is possible to regard a slope as a system.

5. Use annotated diagrams to suggest how the slopes shown in Figure 6.6 may have developed.

7 Human Activity and the Uses of Rocks

KEY WORDS

Quarrying: extracting rock or mineral ore from underground or surface deposits through open-cast techniques.
Fossil fuels: any fuel resource that is derived from the remains of plants that have been subjected to some of the processes associated with the formation of sedimentary rocks.

All landscapes owe much to the underlying rock types and geological structures, as earlier chapters have shown. In addition to influencing the landscape the underlying geology provides challenges and opportunities for the human communities that occupy any given landscape. In upland areas, or areas with rocks or minerals that can provide economic benefit, the nature of the rocks can have an impact on the way of life that is obvious and visible. In lowland areas the relationships are less clear cut, but may be just as significant.

Resistant rocks, such as igneous rocks, weather slowly and form upland areas where settlement is sparse and agricultural activity is limited to sheep farming. Neighbouring sedimentary layers, such as chalk and clay, can produce very different features. Clay will produce deep rich soils, while the thin soils formed on porous chalk will not yield great agricultural rewards.

1 Challenges and Opportunities of Igneous Rocks

a) Landscape

Intrusive igneous rocks tend to be resistant to erosion and to form upland landscapes.

The granite of Dartmoor forms a moorland environment that poses questions to the human communities that are common to other upland areas formed from granite or other igneous rocks. Some of the issues to do with Dartmoor are discussed in the case study about the area in Chapter 2. There are limited possible land uses in upland granite areas because of the nature of the environment. **Agriculture**, **settlement** and **tourism** are human issues that are significantly influenced by the rocks underlying areas such as Dartmoor.

Agriculture is restricted to pastoral farming. Granite is a largely impermeable rock, and the only route water can take through it is down the narrow joints created on cooling. Consequently, granite areas have high drainage densities and poor draining soils. The natural vegetation on Dartmoor is moorland, although there are patches of dwarf oak woodland, such as Wistman's Wood, which might once have covered much of the moor. As a result of the exposed nature of the area and the high drainage densities the only agricultural practices that can be sustained on the moor are pastoral. Sheep are allowed to roam freely across great tracts of the moor and they maintain a distinctive close cropped look to the grasses in areas they visit. On the fringes of the moor, where land is enclosed, some cattle farming occurs, but the open and unfenced nature of the upper areas is not appropriate for cattle farming.

Tourism is focused on outdoor activities. The barren and uninhabited landscape dissected in places by fertile river valleys brings tourists flocking to Dartmoor. The attraction is in being in a rural wilderness area. That attraction has been created by the nature of the granite. As has been discussed, this landscape makes good walking terrain; the views are expansive and tors make good destinations for walkers. As a result Dartmoor's villages can become very busy and overcrowded during the summer season, although those villages will often be very quiet and peaceful during the winter months. It is the draw of wilderness that brings many tourists to Dartmoor each year, and the presence of tors adds to that sense of wilderness because of their unusual craggy appearance.

Dartmoor is not a place blessed with many good sites for **settlement**. The exposed nature of the moor limits shelter, and the carrying capacity of the land is not high. Princetown, in the centre of the moor, is noted mainly for the presence of Dartmoor prison, a forbidding Victorian maximum security institution.

Despite its unpromising look, Dartmoor has been used by people for many years, and it is possible that more people lived high up on the moor than is now the case. A short distance south-east of Houndtor are the remains of a deserted medieval hamlet. This hamlet was set on an east-facing slope overlooking the deep but gentle wooded valley of Becka Brook. To the north rises the huge fractured outcrop of Houndtor, while southward lies the less extensive though equally dominating cliff of Greater Rocks. The substantial remains now visible are those of a cluster of 11 stone buildings – three dwellings and their ancillary structures – that represent the final phase of settlement at Houndtor, from about 1200 to 1350. Excavation in the 1960s revealed evidence that beneath the stone buildings lay a series of turf-walled dwellings dating back to the tenth or eleventh century. The outlying medieval farmhouse to the north was built within a prehistoric enclosure. The whole area was reoccupied in medieval times by the farmers of the village and its outlying homesteads. The inhabitants of

Houndtor would have practised a mixed agricultural economy, growing such crops as corn, oats and rye. Further evidence of cereal growing comes from the presence of kilns built into three of the outbuildings on the main site, which were used to dry corn. It is thought that settlement has died out on Dartmoor because climatic conditions have deteriorated.

b) Minerals

Minerals are also found in igneous rocks. Magma is a very complex mixture of minerals that crystallise at different temperatures. As a result deposits of any given metallic mineral will form in the same place because of the temperature gradient within the intrusion. The Bushveld complex in South Africa is the world's richest deposit of metallic minerals, displaying layers of platinum and chromium, amongst other valuable ores. The layers were segregated during the cooling and crystallising process. The Bushveld complex is extremely ancient, dating back to Precambrian times, and it formed as an intrusive landform. Equally important mineral deposits are found bordering subduction zones when continental crust has melted to create magma. There are vast deposits of copper in Chile associated with the rocks forced into the Andes by the subduction of the Pacific plate. Similarly, Galeras, a volcano in the Columbian Andes, throws half a kilogram of gold into the air each day during its regular eruptions. Accessing these minerals is never straightforward. They are often deposited in narrow **veins** running through the rock. In some areas, such as the Bushveld, open-cast mining is used, but this is environmentally very damaging. In Britain open-cast mining is rarely used, not least because of the high levels of population and stringent planning laws that apply. Quarrying and mining have always been sensitive issues on and around Dartmoor, despite having a long history in the area. Like settlement, there was, perhaps, a greater amount of activity in medieval times than now on the higher parts of the more. Tin, copper and arsenic have been extracted from the moor. It is thought that tin was first mined on Dartmoor, near the River Plym, as long ago as 1168, and there are remains of thirteenth century tin smelting works on the moor.

A deposit of tungsten was discovered in 1867 at Hemerdon, near Plympton. Since then sporadic activity has taken place on site. In 1986 planning permission was given to develop a 200 m deep pit, 850 m by 540 m. It was expected that the mine would have a life of about 20 years. However, the price of tungsten on the world market has not warranted any further working of the deposit.

The granite itself has also been quarried for many years for use as a building stone. With the population explosion associated with the Industrial Revolution and the growth of cities such as Plymouth and Exeter commercial granite quarrying began. Haytor quarry was active by 1820 and granite was removed by a primitive railway and canal. Rock from the Haytor quarry was used to rebuild London Bridge and construct the British Museum. Interestingly, the value of the tors on Dartmoor was understood during the nineteenth century. In 1847 marker stones were placed around Pew Tor to prevent granite being taken from that feature. This pattern was repeated around the moor during the remainder of the century, protecting the natural landscape for future generations.

2 Sedimentary Rocks and Human Activity

Sedimentary landscapes are perhaps the most varied when considering the human activity that goes on within them. While igneous and metamorphic areas tend to be upland areas with sparse settlements and little arable agriculture, sedimentary landscapes provide a wide range of arable and pastoral agriculture, although the landscape often provides more obvious clues about the underlying geology, such as cuesta scenery.

a) Settlement patterns and field boundaries in chalk and clay cuesta scenery

The location of settlements in long inhabited areas owes much to the rock type and structure in those areas. The availability of water was a crucial consideration, and issues such as access to good and varied farmland and wood (for building materials and fuel) were also important considerations. As a result of these issues there is a clear settlement pattern visible in many cuesta areas, such as the Chilterns and the North and South Downs. Settlements developed along the foot of the scarp slope along the **spring line** (Figure 7.1), which shows up on a map as a line of settlements along the foot of the scarp slope of the cuesta. The spring line is created by the intersection of the water table and the impermeable underlying clay. The water sinks through the chalk until it reaches the obstacle posed by the clay, and then moves laterally until it emerges at the foot of the steep slope. This provided a reliable source of freshwater to the inhabitants of the area. The cuesta scenery also provided the varied surroundings that were advantageous. The chalk would have been covered with woodland, probably beech, which could be used as a source of fuel and as building materials, while the clay provides flat land for arable agriculture. Once cleared, the chalk also provided good foraging for pigs and grazing for cattle and sheep.

Field boundaries also vary between sedimentary rocks in chalk and clay scenery. Chalk produces thin, relatively infertile, soils. As a result chalk downlands have seldom been used for arable agriculture. More commonly, they were used for common grazing of livestock, in

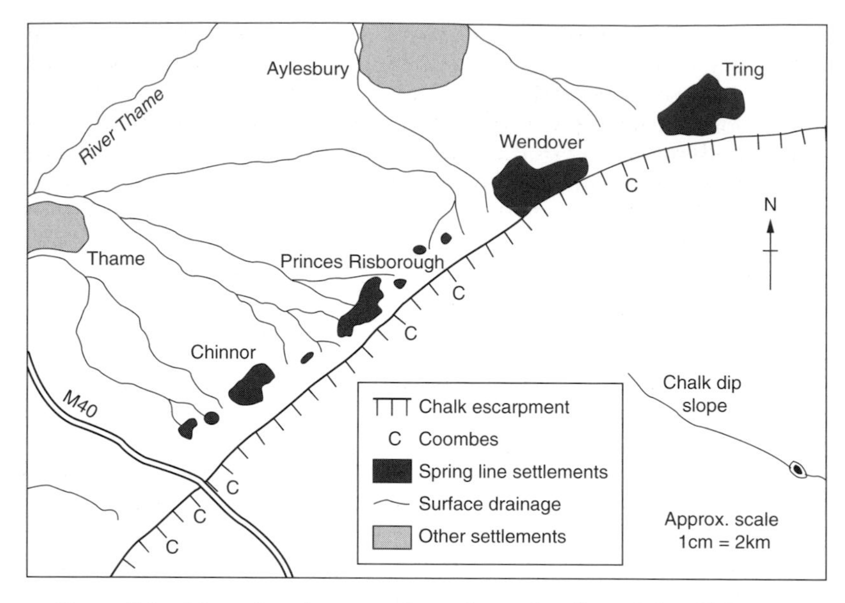

Figure 7.1 Map of settlements along the spring line of the Chilterns, Buckinghamshire.

a way that did not require the land to be enclosed. Consequently, field boundaries on chalk lands are relatively recent innovations and are usually marked by fences. Clay lands, on the other hand, were typically used for arable farming, and were more valuable as a result. This led to greater demands for ownership of the land, and an associated need for a means to determine the location of the field boundaries. As a result of the moist, nutrient-rich nature of the soil hedges grew quickly and were a natural choice for this. The result of these contrasting situations is that patterns of field boundaries can change within a small area.

b) Agriculture and quarrying in Carboniferous limestone areas

Not all sedimentary rocks produce lowland areas. Carboniferous limestone is a resistant rock that produces some scenically very attractive areas, as discussed in Chapter 2. It is also the single most common rock in the British Isles, although this statistic is helped by the great quantities of the rock that are to be found in Ireland. People have always found uses for Carboniferous limestone areas. The caves that develop through solution were used by early inhabitants as a source of shelter as much as 7000 years ago, according to radiometric dating of remains.

Agriculture is, as with other areas, dependent on the soils that form on the underlying calcareous rocks. The soil type found on limestone is a **rendzina**, a thin alkaline soil. A rendzina is an example of a soil that is not climatically determined, but is the product of the rock type below. Soils of this nature are known as **intrazonal soils**. These soils are not especially fertile, but they do yield good grasslands that are ideal grazing territory for sheep. The Yorkshire Dales exemplify this characteristic of Carboniferous limestone areas. The upland area is dissected by valleys that provide water, sufficient flat land for farms to be constructed and clear transport routes. The high ground would be too cold and wet to allow arable agriculture even if the rendzina soils were fertile enough, and so there is no competition for the flocks of sheep. As in the chalk and clay lands, the field boundaries in the landscape are a product of the geology. Dry stone walls are made from loose rocks and boulders found in the local scenery, and are combined skilfully to provide walls up to 2 m high to prevent the sheep leaving their owner's land.

Quarrying in Carboniferous limestone is a major part of the economy of all such areas. Limestone, or the calcite that can be extracted from it, is used for a huge range of purposes. It is used as fertiliser in agriculture, as a catalyst in the making of steel and glass, as a filler in animal feed, paper, paint and pharmaceuticals, and as aggregate in construction. Large boulders are used for rock armour in sea defences. Because Carboniferous limestone areas are also areas of great scenic beauty the existence, work and possible future development of quarries is a very sensitive issue, discussed in the case study.

c) Fossil fuels

Some of the most important resources known to man are produced as part of the group of sedimentary rocks. Coal, oil and natural gas are derived from the partially decomposed remains of ancient organisms preserved within sedimentary rocks. **Coal** is a rock-like substance that is composed of the altered remains of ancient plants that thrived in freshwater swamps. These swamps were located in protected sections of coastal plains, estuaries lagoons and deltas on ancient land masses, where oxygen was not readily available underwater. These conditions could be found over much of what is now central and western Europe during the Carboniferous period some 300–350 million years ago. Swamps were repeatedly created and destroyed across this area, and so layer upon layer of coal has been created in between sandstones produced from fluvial sediments. The dead plant matter was partially decomposed to form **peat**, a forerunner of coal. The burial of the peat beneath more sediment began the processes of diagenesis discussed earlier. These processes are shown in Figure 7.2.

Anthracite is carbon-rich coal formed when coal is subjected to the greater pressures and stresses that could be created by moderately folded and compressed rock layers. The coalfields of South Wales were rich in anthracite, a type of coal that is particularly rich in carbon and burns with fierce heat and creates little pollution.

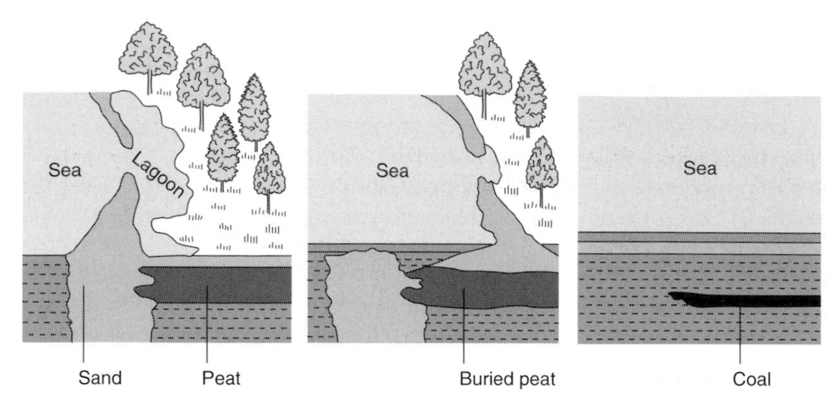

Figure 7.2 The formation of coal.

Oil is produced in shallow basins close to continental areas. The sheltered Gulf of Mexico is thought to be the modern environment most likely to be an oil-forming environment. Organic matter is preserved in oxygen-poor conditions where decay will be at a minimum, and mud accumulates around them. The organic matter then decays into hydrocarbon compounds, a process which is hastened by the compaction created by the addition of layers of sediment. Under pressure the oil will separate into dense and lighter compounds. This mirrors the processes carried out in an oil refinery, although in a much less precise manner.

The oil and gas, along with any water present, will be forced out of the rocks it was formed in by the increasing pressure, and it will concentrate in a **reservoir rock**, one with large, continuous pore spaces that can accommodate the liquid. Sandstones, fragmented limestones and reef deposits can form good reservoir rocks, and if they are surrounded by impermeable rocks the oil and gas will be trapped. Once in the reservoir rock the oil and gas will be driven to an area of low pressure, forming a pool of oil known as an **oil trap**. This process takes millions of years to occur, a stark contrast to the few decades it can take to empty an oil field.

Two typical oil traps are shown in Figure 7.3. The oil and gas has been forced to the area of the reservoir rock where pressure is lowest, for example at the top of the anticline. There are many other geological situations that can form an oil trap and the identification of likely locations for an oil field is a complex business.

3 Metamorphic Rocks and Human Activity

Metamorphic rocks provide no real pattern of landscape, as discussed earlier, and the same is true to some extent of their impact on human activity. Metamorphic rocks tend to be impermeable, and therefore

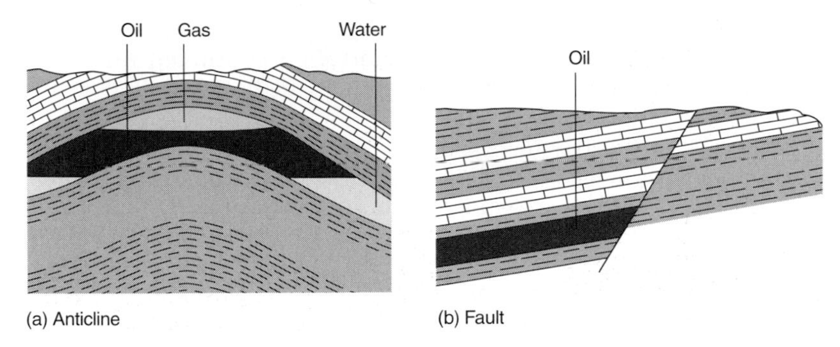

Figure 7.3 Two typical oil traps. *Source*: Dolgoff (1996).

produce landscapes with high drainage densities and flood risks. This means that arable agriculture is not viable in many metamorphic landscapes, and that consequently settlement is often limited to dispersed farms.

However, metamorphic processes have produced a number of valuable mineral resources. Two of the most important products are **kaolin** and **marble**.

a) Kaolin

Kaolin, or **china clay**, is a product of the chemical breakdown of feldspar in granite, a process known as kaolinisation. It is the product of either hydrolysis or **pneumatolysis**, a metamorphic process involving the circulation of hot gases. These processes can affect the granite or the country rock into which it is being intruded. In the latter case this can be said to be a metamorphosis. The feldspar minerals comprising some 30–40% of the granite are decomposed forming the white clay, kaolin.

The product of this process is the UK's second most valuable geological export after petroleum and only salt and gypsum are produced in greater quantities. It has a number of uses. As china clay it is used in the manufacture of glossy paper and porcelain, while as fire clay it is used in the production of pipes and tiles that need to withstand high temperatures.

The major deposits of kaolin in Britain are on the fringes of the south-western granite batholith. Dartmoor and Bodmin Moor have the greatest quantity, and the town of St Austell is the centre of the British kaolin industry. The major area of kaolinisation associated with Dartmoor is around Lee Moor on the southern edge of the granite as shown on Figure 2.7, but many smaller areas of kaolinisation occur. Whiteworks for example, just south-east of Princetown, as the name suggests, is an area where the granite has been kaolinised. This process also affected joints and faults on smaller areas of granite, across the moor. The kaolinisation process probably began once the magma of the intrusion had cooled and then continued for a considerable length of time as heat continued to be generated in the granite by its natural radioactivity.

At Lee Moor there is an extensive complex of china clay workings. The works have been in existence for over 100 years, during which time the workings have expanded with the development of improved mining equipment. The china clay is exported nationally and internationally. The kaolin is extracted through **deep open cut extraction**. As Woodcock (1994) said "surface mineral working is now the most disruptive form of extraction in the British Isles." Deep open cut extraction is the method of mining that creates the greatest scarring to the landscape. In addition to the hole in the ground the unwanted debris, known as the **overburden**, is piled up next to the workings. The overburden from the Lee Moor quarry consists of large piles of mica sand. This sand is of an excellent quality for use in concrete but the cost of transport has so far prohibited significant use. Around ten times as much waste is produced as useable kaolin. Although there are numerous unused pits in the area they are not used to house the overburden from working pits because the unused ones would then be rendered unworkable in the future, should they become economically viable again. Quarry owners now put significant efforts into landscaping their pits, and regularly shaped mounds of waste are vegetated to disguise the worst of the visual pollution. Some disused workings are put to other uses. The famous **Eden Project** is located in a redundant china clay pit near St Austell.

Marble is a metamorphic rock formed from limestone that has been subjected to contact or regional metamorphism. Characteristically it has veins of colour running through it and will take a high polish. As a result it is a popular decorative building stone and many important buildings contain some marble, either structurally or decoratively.

Marble is found in many areas that have once experienced metamorphic activity. Italy is one of Europe's largest marble producers, because of the metamorphic activity caused by the building of the Alps. Italian marble occurs in many localities, from the Alps to Sicily, and is quarried at hundreds of operations. The most important geographic area producing white marble is in the Apuan Alps in Tuscany, particularly near the town of Carrara. The Lazio region, Lombardy, the Po Valley, Puglia, the island of Sicily and the Veneto are important coloured marble-producing areas. About one-half of production was in block form, and 45% of total production was exported. Reserves of Italian marble are considered to be unlimited.

Perhaps the most significant marble production area in the British Isles is Connemara in Ireland.

On the Atlantic coast of County Galway are 12 mountains known as the Twelve Bens. In the centre of this range are the quarries that produce the green marble characteristic of the area. Geologists believe that the sedimentary rocks that now form the marble of the area were laid down some 750 million years ago, and were subsequently metamorphosed to form the marbles that are present today. The Connemara marble deposits are home to several small quarrying firms who use open-cast mining strategies to extract the marble. The highly resistant rock has to be cut with diamond wires before being extracted with a crane.

4 Subsidence

Subsidence is a hazard associated with rocks and geology that is not limited to any particular rock type or any specific environmental circumstance. It can affect any area, and can be caused by natural processes or human activity.

Naturally occurring subsidence can be caused by the deterioration of permafrost in tundra areas, or by solution processes in limestone landscapes.

Human-induced subsidence can be caused by the removal of subterranean material such as oil, gas or coal through extraction, by the alteration of permafrost layers or by disruption of the groundwater system through irrigation or drainage.

Subsidence is usually a gentle process that occurs over a number of years, but occasionally it can be dramatic and catastrophic. Most human-induced subsidence is caused by mining. Gold mining in the Far West Rand of the South African Transvaal has required that the local water table be lowered by 300 m. This has meant that clay-rich material has dried out and shrunk, causing caves to collapse. Consequently, large depressions have developed on the surface, causing buildings to collapse in urban areas.

Oil extraction in the Wilmington oilfield, near Los Angeles, led to the development of 9 m of subsidence between 1928 and 1971, while groundwater extraction around Mexico City has caused 7.5 m of subsidence.

CASE STUDY: QUARRYING IN THE YORKSHIRE DALES

The Carboniferous limestone area of the Yorkshire Dales is an area that provides some of Britain's most scenic upland areas, but it also provides a valuable resource. Limestone is in demand for a number of reasons outlined earlier in this chapter. However, the extraction of the valuable rock is subject to significant controversy, not least because there are several quarries within the Yorkshire Dales National Park, a protected area. National Parks exist to promote and protect the natural environment and beautiful scenery. Quarries scar that scenery and produce dust and noise in an otherwise peaceful location. However, the quarries also provide valuable employment in rural areas where there are often very few job opportunities.

Figure 7.4 shows the distribution of the Great Scar Limestone within the National Park and the location of the major quarries. Quarrying has been carried out in the area for hundreds of years. Stone from small-scale local pits was used to create the Dales scenery of dry stone walls, barns and villages. However, the scale of the industry today is far removed from that of the Middle Ages. In 1998, 4.4 million tonnes of limestone was quarried within the National Park. As Figure 7.4 shows it is not only limestone that is quarried in the National Park; **sandstone** and **gritstones** are also extracted. These are other types of sedimentary rock that are used in the construction industry.

Figure 7.4 Geological map of the Yorkshire Dales National Park, showing quarries.

Quarrying in the Yorkshire Dales is an emotive issue. The arguments usually rehearsed by those in favour and those against are below.

Advantages of quarrying

The Yorkshire Dales contain a living community that has had quarrying as part of its economy for centuries. The industry currently employs 7% of the working population of the Dales, contributing in excess of £7 million per year to the local economy. Dales quarries supply primarily local markets, normally within 80 km of the quarries. If the local quarries were to close the cost and environmental impact of construction in the area would increase dramatically as materials would have to be imported from further afield.

Disadvantages of quarrying

Although quarrying has taken place in the Yorkshire Dales for many years, the modern large-scale quarries are out of keeping with the landscape. The scarring to the landscape is damaging it for the present generation, but also for the future generations. Demand for stone will never stop, and so the demand to quarry in the Dales will continue. In addition to the impact on the natural environment, there is a significant impact on the human communities. The quarrying operations produce noise and dust pollution through the action of the machinery. The transport of the rock on narrow local roads causes traffic jams and increases the risk of accidents. As new permissions are sought it is important that the needs of future locals must be considered, and the landscape must be protected.

Permission to quarry is not granted in perpetuity. Most quarries have to re-apply for permission to quarry every 15 years, and any expansion to a quarry also has to be approved by the authorities. Approval will only be given if the activities meet the requirements of the Environment Act 1995 and the Yorkshire Dales Minerals and Waste Local Plan 1998.

Typically, permission to continue the work of, or expand, a quarry is granted only if the company concerned ensures high standards of environmental control and sensitivity, and has a restoration plan.

Quarries in the Yorkshire Dales have met these requirements in a number of ways:

- **Ingleton quarry** has resited the quarry plant (the sorting and processing machinery usually found within the quarry) so it is less intrusive. Screening banks have been built and planted with trees. A restoration scheme based on a central lake has been approved.

- **Horton quarry** has very large permitted reserves and is very prominent indeed. Major screening and landscaping works have been carried out. The same is true of **Threshfield quarry**.
- **Cool Scar quarry** reached the end of its economically viable reserves and closed in 1999. Restoration for nature conservation has been carried out and completed.
- **Swinden quarry** has upgraded its rail facilities. This keeps 33 000 lorries off the local roads each year. Permission was granted in 1996 to deepen the quarry by 100 m, conditional on the construction of a new plant within the quarry and the removal and landscaping of the old plant. The quarry will close in 2020 after the new workings are exhausted and then restored and managed for nature conservation.

CASE STUDY: PUMICE QUARRYING ON LIPARI

Figure 7.5 Pumice quarry on Lipari.

Pumice is an igneous rock that is ejected from volcanoes. It has a cellular structure that occurs because it forms from the bubbly scum on the surface of a lava lake. Thus, volcanoes that have had lava lakes in their craters often have layers of pumice in the deposits formed by eruptions. Pumice has many uses. It is used in the manufacture of clothing, where it is used in the washing of clothes, and it forms part of lightweight concrete blocks. Its structure means that it is very useful in the construction of insulating materials, both for heat and sound. Most well known, however, is its cosmetic use. Its abrasive surface makes it ideal for the smoothing of dead skin, and it can be found in most high-street chemists. The highest quality of pumice contains high percentages of silica.

It is found in small deposits in many areas with volcanic activity. Lombok, in Indonesia, is a main source, but there is also pumice in areas of the western USA and on some Greek islands.

The Aeolian Islands, off the north coast of Sicily, are active volcanic islands that formed an important part of ancient Roman history. Lipari is one of those islands, and forms the centre of the Italian pumice industry. Pumice has been quarried on Lipari island for a long time; pumice gravel from Lipari was used in the construction of roads and buildings in ancient Rome. Today, around 750 000 tonnes of pumice are quarried on Lipari each year, mainly for use in the European market.

Working in the glaring white quarries was dangerous to eyes and lungs. Before the use of modern breathing protection equipment, many workers died of 'liparosis' (the clogging of the lungs with pumice dust). Doctors at the time remarked they could sharpen their scalpels with the lung tissue of deceased workers.

Summary Diagram

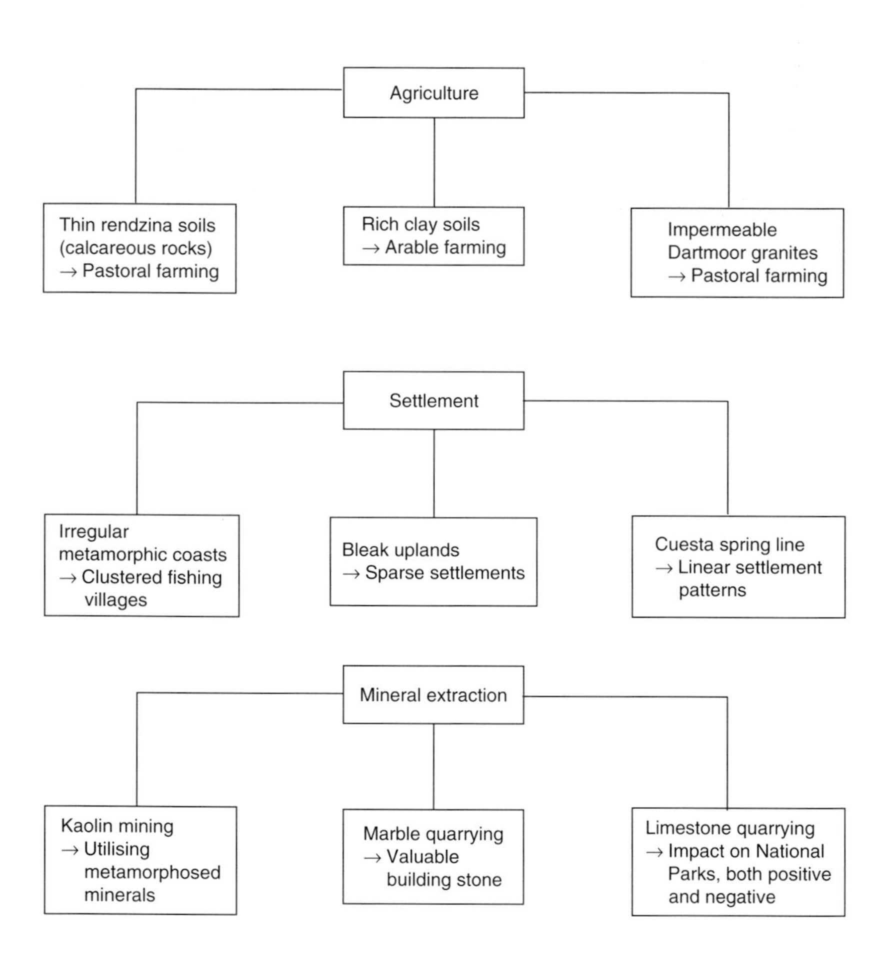

Questions

1. Explain how different geologies can influence the following aspects of human geography:
   a) settlement
   b) agriculture
   c) extractive primary industry.
2. Examine the tourism opportunities that areas of different rock types can offer.
3. Discuss the advantages and disadvantages of quarrying.
4. Why do metamorphic landscapes not provide obvious landscape patterns?
5. Explain the importance of sedimentary rocks in the formation and accessibility of oil.

Bibliography

Atkinson, D.A., 2002, Limestone weathering and features. *Geography Review* 16.

Bishop, V. and Prosser, R., 1997, *Landform Systems*. Collins.

Boot, J., 2003, Glacial erosion in lowland areas. *Geofile* 459.

Bristow, C.M., 1996, *Cornwall's Geology and Scenery: An Introduction*. Cornish Hillside Press.

Caine, N., 1974, The geomorphic processes of the Alpine environment. In: J.D. Ives and R.G. Barry (eds) *Arctic and alpine environments*. Methuen.

Carson, M.A. and Kirby, M.J., 1972, *Hillslope Form and Process*. Cambridge University Press.

Clark, M. and Small, J., 1982, *Slopes and Weathering*. Cambridge University Press.

Collard, R., 1988, *Physical Geography of Landscape*. Collins.

Dalrymple, J.B., Blong, R.J. and Conacher, A.J., 1968, A hypothetical nine-unit land surface model. *Zeit. Geomorph.* 12.

Davis, W.M., 1899, The geographical cycle. *Geographical Journal* 14.

Dolgoff, A., 1996, *Physical Geology*. Heath & Co.

Edwards, D. and King, C., 1999, *Geoscience*. Hodder & Stoughton.

Fortey, R., 1993, *The Hidden Landscape*. Pimlico.

Goudie, A., 1993, *The Nature of the Environment*. Blackwell.

Goudie, A. and Viles, H., 1997, *The Earth Transformed*. Blackwell.

Gudmundsson, A.T. and Kjartansson, H., 1996, *Earth in Action*. Vaka-Helgafell.

Hill, M., 2002, *Arid and Semi-arid Environments*. Hodder & Stoughton.

King, L.C., 1957, The uniformitarian nature of hillslopes. *Transactions of the Edinburgh Geological Society* 17.

Lenon, B. and Cleves, P., 1984, *Fieldwork Techniques and Projects in Geography*. Collins.

Linton, D., 1955, The problem of tors. *Geographical Journal* 121.

Linton, D., 1963, The forms of glacial erosion. *Transactions of the Institute of British Geographers* 33.

McKirdy, A. and Crofts, R., 1999, *Scotland: The Creation of its Natural Landscape*. British Geological Survey.

Mikkelsen, N., Carver, S., Woodward, J. and Christiansen, H.H., 2001, Periglacial processes in the Mestersvig region, central East Greenland. *Geology of Greenland Survey Bulletin* 189.

Nagle, G., 1998, *Hazards*. Nelson.

Nagle, G., 2000, *Advanced Geography*. Oxford University Press.

Palmer, J. and Neilson, R., 1962, The origin of granite tors on Dartmoor, Devonshire. *Proceedings of the Yorkshire Geological Society* 33.

Penck, W., 1924, *Die morphologische analyse*. Stuttgart.

Prosser, R., 1992, *Natural Systems and Human Responses*. Nelson.

Punnett, N., 2003, Granite landscapes: Dartmoor. *Geofile* 436.

Searle, M., 2003, High and rising. *Geographical Magazine* May 2003.

Selby, M.J., 1985, *Earth's Changing Surface*. Oxford University Press.

Strahler, A., 2003, *Introducing Physical Geography*. John Wiley.

Summerfield, M., 1991, *Global Geomorphology*. Longman.

Varnes D.J., 1978, Slope movement and types and processes. In: R.L. Schuster and R.J. Krizek (eds) *Landslides: Analysis and Control.* Transportation Research Board Special Report 176, National Academy of Sciences, Washington DC.
Waugh, D., 2002, *Geography. An Integrated Approach* Nelson Thomas.
Whittow, J., 1984, *Dictionary of Physical Geography*. Penguin.
Woodcock, N., 1994, *Geology and Environment in Britain and Ireland.* Routledge.

Websites

A Dartmoor village: www.manaton.org.uk
Himalyan tectonics: http://earth.leeds.ac.uk/dynamicearth/himalayas
Human-induced mass movement: www.aber.ac.uk/iges/cti-g/hazards2000/massmovement/human2.html
St Helena: www.btinternet.com/~sa_sa/st_helena/st_helena_geology.html

Index